Fundamentals of Crew Leadership

Trainee Guide

Pearson

Boston Columbus Indianapolis New York San Francisco Amsterdam
Cape Town Dubai London Madrid Milan Munich Paris Montreal Toronto Delhi
Mexico City Sao Paulo Sydney Hong Kong Seoul Singapore Taipei Tokyo

NCCER

President and Chief Executive Officer: Don Whyte
President: Boyd Worsham
Vice President: Steve Greene
Chief Operations Officer: Katrina Kersch
Fundamentals of Crew Leadership Project Manager: Mark Thomas
Senior Development Manager: Mark Thomas
Senior Production Manager: Tim Davis
Art *Manager:* Kelly Sadler
Managing Editor: Natalie Richoux
Desktop Publishing Manager: James McKay
Permissions Specialist: Adrienne Payne
Production Specialists: Gene Page, Eric Caraballoso
Editors: Graham Hack, Jordan Hutchinson

Writing and development services provided by Topaz Publications, Liverpool, NY

Lead Writer/Project Manager: Thomas Burke
Desktop Publisher: Joanne Hart
Art Director: Alison Richmond
Permissions Editor: Andrea LaBarge
Writer: Thomas Burke

Pearson

Director of Alliance/Partnership Management: Kelly Trakalo
Content Producer: Alexandrina B. Wolf
Assistant Content Producer: Alma Dabral
Digital Content Producer: Jose Carchi
Senior Marketing Manager: Brian Hoehl
Composition: NCCER
Printer/Binder: LSC Communications
Cover Printer: LSC Communications
Text Fonts: Palatino and Univers

Credits and acknowledgments for content borrowed from other sources and reproduced, with permission, in this textbook appear at the end of each module.

15 2022

Pearson

ISBN-13: 978-0-13-487188-2
ISBN-10: 0-13-487188-X

Preface

To the Trainee

In every endeavor, leaders must understand the structure and mission of their organizations. They must also internalize and competently apply the techniques of professional leadership and management to communicate with their subordinates. Leaders must make effective decisions to solve problems, creating a motivated team capable of accomplishing its part of the organization's mission.

The crew leader is the first level of supervision for most craft workers. As a crew leader, you are responsible for completion of the job within specifications. You are also responsible for ensuring the safety, well-being, training, and professional development of your crew members. Effective crew leaders understand and apply the basics of planning, scheduling, and cost control while motivating crew members to work safely to complete job tasks at the required level of quality. As the best possible crew leader, you will actively train and mentor your crew members, ensuring that a qualified leader is ready to move into crew leadership when you are ready for promotion to a higher level of responsibility and authority in your organization.

We wish you success as you embark on your journey into leadership and management and hope that you will continue your training beyond this textbook. There are more than 6 million construction craft professionals in the United States, and there are many opportunities awaiting those with the skills and desire to move forward in the construction industry. As a leader within the construction industry, your job is now to use all of the tools at your command to ensure that your team accomplishes its part of your organization's mission.

Following *Fundamentals of Crew Leadership*, the NCCER management training module *Mentoring for Craft Professionals* (module ID 46102) will teach you the skills you need to effectively guide the development of your team members into craft professionals of the high caliber expected and needed within the construction industry. As you gain experience and knowledge within your craft, continue your leadership and management training with *Project Supervision*, *Project Management*, and *Construction Workforce Development Professional* in order to be prepared to lead at every level and develop tomorrow's workforce of craft professionals.

New with *Fundamentals of Crew Leadership*

NCCER and Pearson are pleased to present *Fundamentals of Crew Leadership*. This edition has been updated to streamline the presentation of training and includes revisions to the module examinations.

Fundamentals of Crew Leadership is now presented in NCCER's most up-to-date instructional systems design, in which the content of each module is tied directly to a learning objective. There are additional study questions at the end of each section of text. Images and diagrams have been updated throughout the program to reflect current electrical practices with an emphasis on safety. This new version of NCCER's management training module "Fundamentals of Crew Leadership" (module ID 46101) is available as a standalone training module, and will also be part of many other NCCER craft training curricula beginning with the current edition of *Electrical Level Four*.

We wish you success as you progress through this training program. If you have any comments on how NCCER might improve upon this textbook, please complete the User Update form located at the back of each module and send it to us. We will always consider and respond to input from our customers.

We invite you to visit the NCCER website at **www.nccer.org** for information on the latest product releases and training, as well as online versions of the *Cornerstone* magazine and Pearson's NCCER product catalog.

Your feedback is welcome. You may email your comments to **curriculum@nccer.org** or send general comments and inquiries to **info@nccer.org**.

NCCER Standardized Curricula

NCCER is a not-for-profit 501(c)(3) education foundation established in 1996 by the world's largest and most progressive construction companies and national construction associations. It was founded to address the severe workforce shortage facing the industry and to develop a standardized training process and curricula. Today, NCCER is supported by hundreds of leading construction and maintenance companies, manufacturers, and national associations. The NCCER Standardized Curricula was developed by NCCER in partnership with Pearson, the world's largest educational publisher.

Some features of the NCCER Standardized Curricula are as follows:

- An industry-proven record of success
- Curricula developed by the industry, for the industry
- National standardization providing portability of learned job skills and educational credits
- Compliance with the Office of Apprenticeship requirements for related classroom training (*CFR 29:29*)
- Well-illustrated, up-to-date, and practical information

NCCER also maintains the NCCER Registry, which provides transcripts, certificates, and wallet cards to individuals who have successfully completed a level of training within a craft in NCCER's Curricula. *Training programs must be delivered by an NCCER Accredited Training Sponsor in order to receive these credentials.*

Special Features

In an effort to provide a comprehensive and user-friendly training resource, this curriculum showcases several informative features. Whether you are a visual or hands-on learner, these features are intended to enhance your knowledge of the construction industry as you progress in your training. Some of the features you may find in the curriculum are explained below.

Introduction

This introductory page, found at the beginning of each module, lists the module Objectives, Performance Tasks, and Trade Terms. The Objectives list the knowledge you will acquire after successfully completing the module. The Performance Tasks give you an opportunity to apply your knowledge to real-world tasks. The Trade Terms are industry-specific vocabulary that you will learn as you study this module.

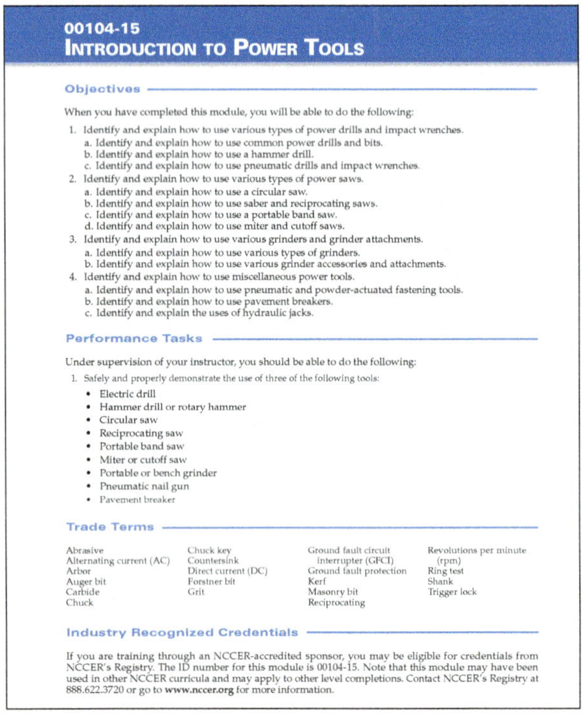

00104-15
INTRODUCTION TO POWER TOOLS

Objectives

When you have completed this module, you will be able to do the following:

1. Identify and explain how to use various types of power drills and impact wrenches.
   a. Identify and explain how to use common power drills and bits.
   b. Identify and explain how to use a hammer drill.
   c. Identify and explain how to use pneumatic drills and impact wrenches.
2. Identify and explain how to use various types of power saws.
   a. Identify and explain how to use a circular saw.
   b. Identify and explain how to use saber and reciprocating saws.
   c. Identify and explain how to use a portable band saw.
   d. Identify and explain how to use miter and cutoff saws.
3. Identify and explain how to use various grinders and grinder attachments.
   a. Identify and explain how to use various types of grinders.
   b. Identify and explain how to use various grinder accessories and attachments.
4. Identify and explain how to use miscellaneous power tools.
   a. Identify and explain how to use pneumatic and powder-actuated fastening tools.
   b. Identify and explain how to use pavement breakers.
   c. Identify and explain the uses of hydraulic jacks.

Performance Tasks

Under supervision of your instructor, you should be able to do the following:

1. Safely and properly demonstrate the use of three of the following tools:
   - Electric drill
   - Hammer drill or rotary hammer
   - Circular saw
   - Reciprocating saw
   - Portable band saw
   - Miter or cutoff saw
   - Portable or bench grinder
   - Pneumatic nail gun
   - Pavement breaker

Trade Terms

Abrasive	Chuck key	Ground fault circuit interrupter (GFCI)	Revolutions per minute (rpm)
Alternating current (AC)	Countersink	Ground fault protection	Ring test
Arbor	Direct current (DC)	Kerf	Shank
Auger bit	Forstner bit	Masonry bit	Trigger lock
Carbide	Grit	Reciprocating	
Chuck			

Industry Recognized Credentials

If you are training through an NCCER-accredited sponsor, you may be eligible for credentials from NCCER's Registry. The ID number for this module is 00104-15. Note that this module may have been used in other NCCER curricula and may apply to other level completions. Contact NCCER's Registry at 888.622.3720 or go to **www.nccer.org** for more information.

Trade Features

Trade features present technical tips and professional practices based on real-life scenarios similar to those you might encounter on the job site.

Bowline Trivia

Some people use this saying to help them remember how to tie a bowline: "The rabbit comes out of his hole, around a tree, and back into the hole."

Figures and Tables

Photographs, drawings, diagrams, and tables are used throughout each module to illustrate important concepts and provide clarity for complex instructions. Text references to figures and tables are emphasized with *italic* type.

Notes, Cautions, and Warnings

Safety features are set off from the main text in highlighted boxes and categorized according to the potential danger involved. Notes simply provide additional information. Cautions flag a hazardous issue that could cause damage to materials or equipment. Warnings stress a potentially dangerous situation that could result in injury or death to workers.

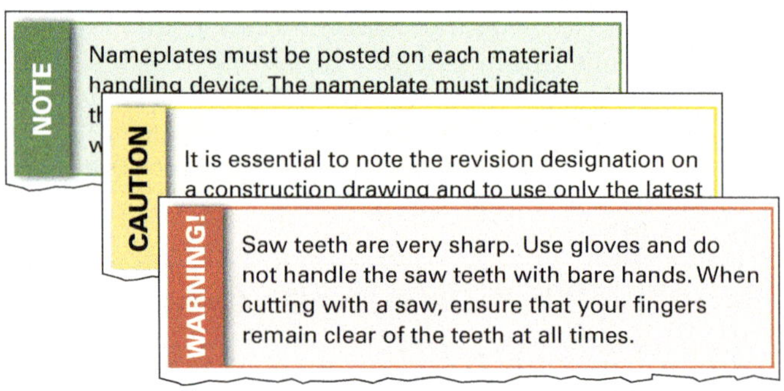

Case History

Case History features emphasize the importance of safety by citing examples of the costly (and often devastating) consequences of ignoring best practices or OSHA regulations.

> **Case History**
>
> **Requesting an Outage**
>
> An electrical contractor requested an outage when asked to install two bolt-in, 240V breakers in panels in a data processing room. It was denied due to the 24/7 worldwide information processing hosted by the facility. The contractor agreed to proceed only if the client would sign a letter agreeing not to hold them responsible if an event occurred that damaged computers or resulted in loss of data. No member of upper management would accept liability for this possibility, and the outage was scheduled.
>
> **The Bottom Line:** If you can communicate the liability associated with an electrical event, you can influence management's decision to work energized.

Going Green

Going Green features present steps being taken within the construction industry to protect the environment and save energy, emphasizing choices that can be made on the job to preserve the health of the planet.

> GOING GREEN
>
> **Reducing Your Carbon Footprint**
>
> Many companies are taking part in the paperless movement. They reduce their environmental impact by reducing the amount of paper they use. Using email helps to reduce the amount of paper used,

Did You Know

Did You Know features introduce historical tidbits or interesting and sometimes surprising facts about the trade.

> **Did You Know?**
>
> **Safety First**
>
> Safety training is required for all activities. Never operate tools, machinery, or equipment without prior training. Always refer to the manufacturer's instructions.

Step-by-Step Instructions

Step-by-step instructions are used throughout to guide you through technical procedures and tasks from start to finish. These steps show you how to perform a task safely and efficiently.

> Perform the following steps to erect this system area scaffold:
>
> *Step 1* Gather and inspect all scaffold equipment for the scaffold arrangement.
>
> *Step 2* Place appropriate mudsills in their approximate locations.
>
> *Step 3* Attach the screw jacks to the mudsills.

Trade Terms

Each module presents a list of Trade Terms that are discussed within the text and defined in the Glossary at the end of the module. These terms are presented in the text with **bold, blue** type upon their first occurrence. To make searches for key information easier, a comprehensive Glossary of Trade Terms from all modules is located at the back of this book.

> During a rigging operation, the **load** being lifted or moved must be connected to the apparatus, such as a crane, that will provide the power for movement. The connector—the link between the load and the apparatus—is often a sling made of synthetic, chain, or **wire rope** materials. This section focuses on three types of slings:

Section Review

Each section of the module wraps up with a list of Additional Resources for further study and Section Review questions designed to test your knowledge of the Objectives for that section.

> **Additional Resources**
>
> *Materials Handling Handbook*, The American Society of Mechanical Engineers (ASME) and The International Material Management Society (IMMS), Raymond A. Kulwiec, Editor-in-Chief. 1985. New York, NY: Wiley-Interscience.
>
> *Manufacturing Facilities Design & Material Handling*, Matthew P. Stevens, Fred E. Meyers. 2013. West Lafayette, IN: Purdue University Press.
>
> **1.0.0 Section Review**
>
> 1. For material handling tasks, it is just as important to be mentally fit as it is to be _____.
>   a. physically fit
>   b. physically aggressive
>   c. closely supervised
>   d. over 200 pounds
>
> 2. Which of the following is a type of knot that is often used to join the ends of two ropes in non-critical, low-strain applications?
>   a. Bowline
>   b. Clove hitch
>   c. Half hitch
>   d. Square knot

Review Questions

The end-of-module Review Questions can be used to measure and reinforce your knowledge of the module's content.

> **Review Questions**
>
> 1. Identification tags for slings must include the _____.
>   a. type of protective pads to use
>   b. type of damage sustained during use
>   c. color of the tattle-tail
>   d. manufacturer's name or trademark
>
> 2. The type of wire rope core that is susceptible to heat damage at relatively low temperatures is the _____.
>   a. fiber core
>   b. strand core
>   c. independent wire rope core
>   d. metallic link supporting core
>
> 3. Synthetic slings must be inspected _____.
>   a. once every month
>   b. visually at the start of each work week
>   c. before every use
>   d. once wear or damage becomes apparent
>
> 4. An alloy steel chain sling must be removed from service if there is evidence that _____.
>   a. the sling has been used in different hitch configurations
>   b. replacement links have been used to repair the chain
>   c. the sling has been used for more than one year
>   d. strands in the supporting core have weakened
>
> 5. A piece of rigging hardware used to couple the end of a wire rope to eye fittings, hooks, or other connections is a(n) _____.
>   a. eyebolt
>   b. hitch
>   c. shackle
>   d. U-bolt
>
> 6. A lifting clamp is most likely to be used to move loads such as _____.
>   a. steel plates
>   b. piping bundles
>   c. concrete blocks
>   d. plastic tubing
>
> 7. Chain hoists are able to lift heavy loads by utilizing a _____.
>   a. rope and pulley system
>   b. rigger's strength
>   c. stationary counterweight
>   d. gear system
>
> 8. Before attempting to lift a load with a chain hoist, make sure that the _____.
>   a. hoist is secured to a come-along
>   b. load is properly balanced
>   c. tag lines are properly anchored
>   d. tackle is connected to its power source
>
> 9. A hitch configuration that allows slings to be connected to the same load without using a spreader beam is a _____.
>   a. double-wrap hitch
>   b. choker hitch
>   c. bridle hitch
>   d. basket hitch
>
> 10. To make the emergency stop signal that is used by riggers, extend both arms _____.
>   a. horizontally with palms down and quickly move both arms back and forth
>   b. directly in front and then move both arms up and down repeatedly
>   c. vertically above the head and wave both arms back and forth
>   d. horizontally with clenched fists and move both arms up and down
>
> 00106-15 Introduction to Basic Rigging Module Six 31

NCCER Standardized Curricula

NCCER's training programs comprise more than 80 construction, maintenance, pipeline, and utility areas and include skills assessments, safety training, and management education.

Boilermaking
Cabinetmaking
Carpentry
Concrete Finishing
Construction Craft Laborer
Construction Technology
Core Curriculum: Introductory Craft Skills
Drywall
Electrical
Electronic Systems Technician
Heating, Ventilating, and Air Conditioning
Heavy Equipment Operations
Heavy Highway Construction
Hydroblasting
Industrial Coating and Lining Application Specialist
Industrial Maintenance Electrical and Instrumentation Technician
Industrial Maintenance Mechanic
Instrumentation
Insulating
Ironworking
Masonry
Millwright
Mobile Crane Operations
Painting
Painting, Industrial
Pipefitting
Pipelayer
Plumbing
Reinforcing Ironwork
Rigging
Scaffolding
Sheet Metal
Signal Person
Site Layout
Sprinkler Fitting
Tower Crane Operator
Welding

Maritime

Maritime Industry Fundamentals
Maritime Pipefitting
Maritime Structural Fitter

Green/Sustainable Construction

Building Auditor
Fundamentals of Weatherization
Introduction to Weatherization
Sustainable Construction Supervisor
Weatherization Crew Chief
Weatherization Technician
Your Role in the Green Environment

Energy

Alternative Energy
Introduction to the Power Industry
Introduction to Solar Photovoltaics
Introduction to Wind Energy
Power Industry Fundamentals
Power Generation Maintenance Electrician
Power Generation I&C Maintenance Technician
Power Generation Maintenance Mechanic
Power Line Worker
Power Line Worker: Distribution
Power Line Worker: Substation
Power Line Worker: Transmission
Solar Photovoltaic Systems Installer
Wind Turbine Maintenance Technician

Pipeline

Control Center Operations, Liquid
Corrosion Control
Electrical and Instrumentation
Field Operations, Liquid
Field Operations, Gas
Maintenance
Mechanical

Safety

Field Safety
Safety Orientation
Safety Technology

Supplemental Titles

Applied Construction Math
Tools for Success

Management

Construction Workforce Development Professional
Fundamentals of Crew Leadership
Mentoring for Craft Professionals
Project Management
Project Supervision

Spanish Titles

Acabado de concreto: nivel uno (*Concrete Finishing Level One*)
Aislamiento: nivel uno (*Insulating Level One*)
Albañilería: nivel uno (*Masonry Level One*)
Andamios (*Scaffolding*)
Carpintería: Formas para carpintería, nivel tres (*Carpentry: Carpentry Forms, Level Three*)
Currículo básico: habilidades introductorias del oficio (*Core Curriculum: Introductory Craft Skills*)
Electricidad: nivel uno (*Electrical Level One*)
Herrería: nivel uno (*Ironworking Level One*)
Herrería de refuerzo: nivel uno (*Reinforcing Ironwork Level One*)
Instalación de rociadores: nivel uno (*Sprinkler Fitting Level One*)
Instalación de tuberías: nivel uno (*Pipefitting Level One*)
Instrumentación: nivel uno, nivel dos, nivel tres, nivel cuatro (*Instrumentation Levels One through Four*)
Orientación de seguridad (*Safety Orientation*)
Paneles de yeso: nivel uno (*Drywall Level One*)
Seguridad de campo (*Field Safety*)

Acknowledgments

This curriculum was revised as a result of the farsightedness and leadership of the following sponsors:

ABC South Texas Chapter
HB Training & Consulting
Turner Industries Group, LLC
University of Georgia
Vision Quest Academy
Willmar Electric Service

This curriculum would not exist were it not for the dedication and unselfish energy of those volunteers who served on the Authoring Team. A sincere thanks is extended to the following:

John Ambrosia
Harold (Hal) Heintz
Mark Hornbuckle
Jonathan Liston
Jay Tornquist
Wayne Tyson
Antonio "Tony" Vazquez

NCCER Partners

American Council for Construction Education
American Fire Sprinkler Association
Associated Builders and Contractors, Inc.
Associated General Contractors of America
Association for Career and Technical Education
Association for Skilled and Technical Sciences
Construction Industry Institute
Construction Users Roundtable
Design Build Institute of America
GSSC – Gulf States Shipbuilders Consortium
ISN
Manufacturing Institute
Mason Contractors Association of America
Merit Contractors Association of Canada
NACE International
National Association of Women in Construction
National Insulation Association
National Technical Honor Society
National Utility Contractors Association
NAWIC Education Foundation
North American Crane Bureau
North American Technician Excellence
Pearson
Prov
SkillsUSA®
Steel Erectors Association of America
U.S. Army Corps of Engineers
University of Florida, M. E. Rinker Sr., School of Construction Management
Women Construction Owners & Executives, USA

NCCER Business Partners

Contents

Fundamentals of Crew Leadership

Covers basic leadership skills and explains different leadership styles, communication, delegating, and problem solving. Job-site safety and the crew leader's role in safety are also discussed. Explains details and procedures for project planning, scheduling, and estimating. (Module ID 46101; 20 Hours)

Index

Fundamentals of Crew Leadership

Overview

When a crew is assembled to complete a job, one person is appointed the leader. This person is usually an experienced craft professional who has demonstrated leadership qualities. While having natural leadership qualities helps in becoming an effective leader, it is more true that "leaders are made, not born." Whether you are a crew leader or want to become one, this module will help you learn more about the requirements and skills needed to succeed.

Module 46101

NCCER

President: Don Whyte
Vice President: Steve Greene
Chief Operations Officer: Katrina Kersch
Fundamentals of Crew Leadership Project Manager: Mark Thomas
Senior Development Manager: Mark Thomas
Senior Production Manager: Tim Davis
Quality Assurance Coordinator: Karyn Payne
Desktop Publishing Coordinator: James McKay
Permissions Specialist: Adrienne Payne
Production Specialist: Adrienne Payne
Editor: Graham Hack

Writing and development services provided by Topaz Publications, Liverpool, NY

Lead Writer/Project Manager: Thomas Burke
Desktop Publisher: Joanne Hart
Art Director: Alison Richmond
Permissions Editor: Andrea LaBarge
Writer: Thomas Burke

Pearson

Director of Alliance/Partnership Management: Andrew Taylor
Editorial Assistant: Collin Lamothe
Program Manager: Alexandrina B. Wolf
Assistant Content Producer: Alma Dabral
Digital Content Producer: Jose Carchi
Director of Marketing: Leigh Ann Simms
Senior Marketing Manager: Brian Hoehl
Composition: NCCER
Printer/Binder: RR Donnelley
Cover Printer: RR Donnelley
Text Fonts: Palatino, Minion Pro, and Univers

Trainees with successful module completions may be eligible for credentialing through the NCCER Registry. To learn more, go to **www.nccer.org** or contact us at 1.888.622.3720. Our website has information on the latest product releases and training, as well as online versions of our *Cornerstone* magazine and Pearson's product catalog.

Your feedback is welcome. You may email your comments to **curriculum@nccer.org**, send general comments and inquiries to **info@nccer.org**, or fill in the User Update form at the back of this module.

This information is general in nature and intended for training purposes only. Actual performance of activities described in this manual requires compliance with all applicable operating, service, maintenance, and safety procedures under the direction of qualified personnel. References in this manual to patented or proprietary devices don't constitute a recommendation of their use.

46101 V3

46101
Fundamentals of Crew Leadership

Objectives

When you have completed this module, you will be able to do the following:

1. Describe current issues and organizational structures in industry today.
   a. Describe the leadership issues facing the construction industry.
   b. Explain how gender and cultural issues affect the construction industry.
   c. Explain the organization of construction businesses and the need for policies and procedures.
2. Explain how to incorporate leadership skills into work habits, including communications, motivation, team-building, problem-solving, and decision-making skills.
   a. Describe the role of a leader on a construction crew.
   b. Explain the importance of written and oral communication skills.
   c. Describe methods for motivating team members.
   d. Explain the importance of teamwork to a construction project.
   e. Identify effective problem-solving and decision-making methods.
3. Identify a crew leader's typical safety responsibilities with respect to common safety issues, including awareness of safety regulations and the cost of accidents.
   a. Explain how a strong safety program can enhance a company's success.
   b. Explain the purpose of OSHA and describe the role of OSHA in administering worker safety.
   c. Describe the role of employers in establishing and administering safety programs.
   d. Explain how crew leaders are involved in administering safety policies and procedures.
4. Demonstrate a basic understanding of the planning process, scheduling, and cost and resource control.
   a. Describe how construction contracts are structured.
   b. Describe the project planning and scheduling processes.
   c. Explain how to implement cost controls on a construction project.
   d. Explain the crew leader's role in controlling project resources and productivity.

Performance Tasks

Under the supervision of your instructor, you should be able to do the following:

1. Develop and present a look-ahead schedule.
2. Develop an estimate for a given work activity.

Trade Terms

Autonomy
Bias
Cloud-based applications
Craft professionals
Crew leader
Critical path
Demographics
Ethics
Infer
Intangible
Job description
Job diary
Legend
Lethargy
Letter of instruction (LOI)
Local area networks (LAN)
Lockout/tagout (LOTO)
Look-ahead schedule
Negligence
Organizational chart
Paraphrase
Pragmatic
Proactive
Project manager
Return on investment (ROI)
Safety data sheets (SDS)
Sexual harassment
Smartphone
Superintendent
Synergy
Textspeak
Wide area networks (WAN)
Wi-Fi
Work breakdown structure (WBS)

Industry Recognized Credentials

If you are training through an NCCER-accredited sponsor, you may be eligible for credentials from NCCER's Registry. The ID number for this module is 46101. Note that this module may have been used in other NCCER curricula and may apply to other level completions. Contact NCCER's Registry at 888.622.3720 or go to **www.nccer.org** for more information.

Contents

1.0.0 Business Structures and Issues in the Industry 1
1.1.0 Leadership Issues and Training Strategies 2
1.1.1 Motivation 2
1.1.2 Understanding Workers 2
1.1.3 Craft Training 3
1.1.4 Supervisory Training 3
1.1.5 Impact of Technology 3
1.2.0 Gender and Cultural Issues 4
1.2.1 Communication Styles of Men and Women 5
1.2.2 Language Barriers 5
1.2.3 Cultural Differences 6
1.2.4 Sexual Harassment 6
1.2.5 Gender and Minority Discrimination 7
1.3.0 Business Organization 8
1.3.1 Division of Responsibility 8
1.3.2 Responsibility, Authority, and Accountability 9
1.3.3 Job Descriptions 9
1.3.4 Policies and Procedures 10
2.0.0 Leadership Skills 13
2.1.0 The Qualities and Role of a Leader 14
2.1.1 Functions of a Leader 15
2.1.2 Leadership Traits 15
2.1.3 Expected Leadership Behavior 15
2.1.4 Leadership Styles 15
2.1.5 Ethics in Leadership 16
2.2.0 Communication 17
2.2.1 The Communication Process and Verbal Communication 17
2.2.2 Written or Visual Communications 19
2.2.3 Nonverbal Communication 19
2.2.4 Communication Issues 20
2.3.0 Motivation 20
2.3.1 Accomplishment 20
2.3.2 Change 22
2.3.3 Job Importance 22
2.3.4 Opportunity for Advancement 22
2.3.5 Recognition 23
2.3.6 Personal Growth 23
2.3.7 Rewards 23
2.3.8 Motivating Employees 24
2.4.0 Team Building 24
2.4.1 Successful Teams 24
2.4.2 Building Successful Teams 24
2.4.3 Delegating 25
2.4.4 Implementing Policies and Procedures 26

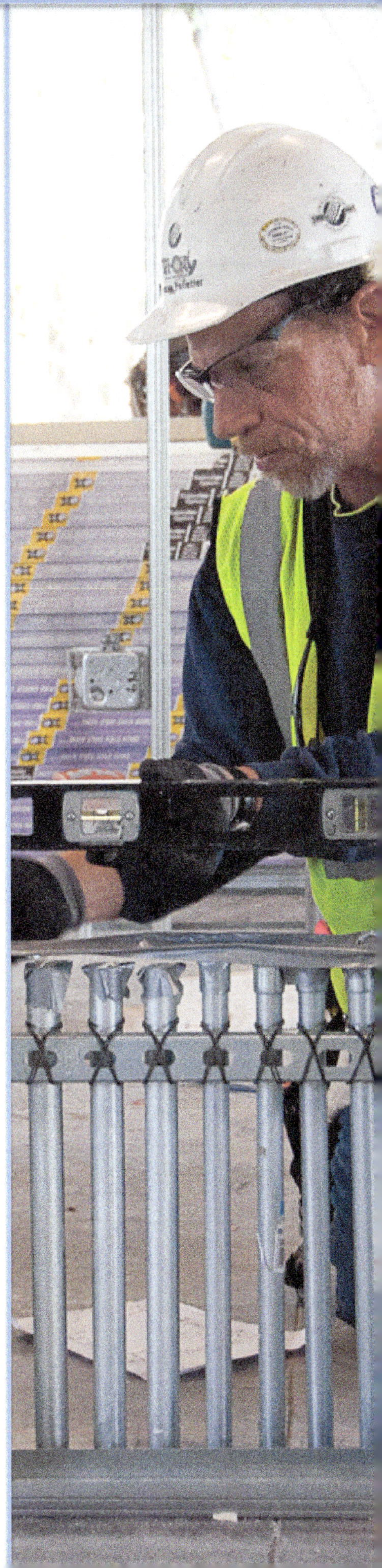

2.5.0 Making Decisions and Solving Problems 27
2.5.1 Decision Making Versus Problem Solving 27
2.5.2 Types of Decisions 27
2.5.3 Problem Solving 27
2.5.4 Special Leadership Problems 28
3.0.0 Safety and Safety Leadership 32
3.1.0 The Impact of Accidents 32
3.1.1 Cost of Accidents 33
3.2.0 OSHA 34
3.3.0 Employer Safety Responsibilities 35
3.3.1 Safety Program 35
3.3.2 Safety Policies and Procedures 35
3.3.3 Hazard Identification and Assessment 36
3.3.4 Safety Information and Training 36
3.3.5 Safety Record System 36
3.3.6 Accident Investigation 37
3.4.0 Leader Involvement in Safety 38
3.4.1 Safety Training Sessions 38
3.4.2 Inspections 39
3.4.3 First Aid 39
3.4.4 Fire Protection and Prevention 40
3.4.5 Substance Abuse 40
3.4.6 Job-Related Accident Investigations 41
3.4.7 Promoting Safety 41
3.4.8 Safety Contests 42
3.4.9 Incentives and Rewards 42
3.4.10 Publicity 43
4.0.0 Project Planning 46
4.1.0 Construction Project Phases, Contracts, and Budgeting 46
4.1.1 Development Phase 46
4.1.2 Planning Phase 47
4.1.3 Construction Phase 48
4.1.4 Project-Delivery Systems 48
4.1.5 Cost Estimating and Budgeting 48
4.2.0 Planning 54
4.2.1 Stages of Planning 57
4.2.2 The Planning Process 58
4.2.3 Planning Resources 60
4.2.4 Scheduling 61
4.3.0 Cost Control 65
4.3.1 Assessing Cost Performance 65
4.3.2 Field Reporting System 65
4.3.3 Crew Leader's Role in Cost Control 67

4.4.0 Resource Control ... 68
4.4.1 Ensuring On-Time Delivery ... 68
4.4.2 Preventing Waste ... 68
4.4.3 Verifying Material Delivery ... 69
4.4.4 Controlling Delivery and Storage ... 69
4.4.5 Preventing Theft and Vandalism ... 70
4.4.6 Equipment Control ... 70
4.4.7 Tool Control ... 70
4.4.8 Labor Control ... 71
4.4.9 Production and Productivity ... 71
Appendix OSHA Forms for Safety Records ...

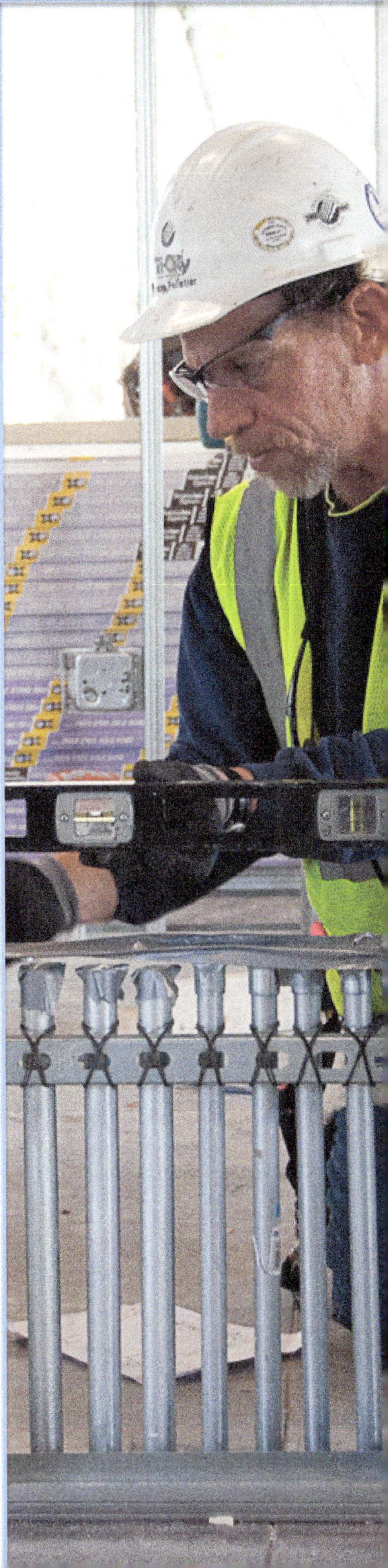

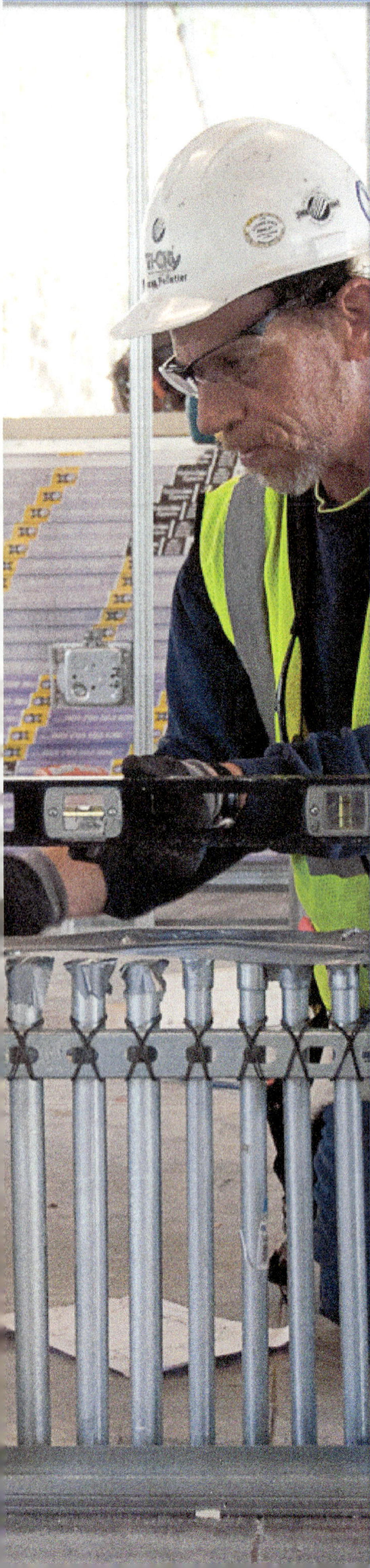

Figures and Tables

Figure 1 Digital tablets are a convenient management and scheduling tool for supervisors 4
Figure 2 The modern workforce is diverse 5
Figure 3 Any unwelcome physical contact can be considered harassment 7
Figure 4 Sample organization chart for a construction company 9
Figure 5 Sample organization chart for an industrial company 10
Figure 6 A sample job description 11
Figure 7 Percentages of technical and supervisory work by role 14
Figure 8 The authority-autonomy leadership style matrix 16
Figure 9 The communication process 17
Figure 10 Eye contact is important when communicating face-to-face 18
Figure 11 Avoid textspeak for on-the-job written communications 19
Figure 12 Tailor your message 23
Figure 13 Falls are the leading cause of deaths and injuries in construction 33
Figure 14 Costs associated with accidents 34
Figure 15 Root causes of accidents 38
Figure 16 An impaired worker is a dangerous one 41
Figure 17 An example of a safety award 43
Figure 18 Architects and clients meet to refine plans 47
Figure 19 Project flow diagram 49
Figure 20 Project-delivery systems 50
Figure 21 Quantity takeoff sheet 52
Figure 22 Summary sheet 53
Figure 23 Proper prior planning prevents poor performance 54
Figure 24 Steps to effective planning 59
Figure 25 Planning form 60
Figure 26 Sample page from a job diary 60
Figure 27 Example of a bar-chart schedule 64
Figure 28 Example of a network schedule 65
Figure 29 Short-term schedule 66
Figure 30 Idle workers cost money 67
Figure 31 Waste materials separated for recycling 69

Table 1 OSHA Penalties for violations established in 2016 35

Section One

1.0.0 Business Structures and Issues in the Industry

Objective

Describe current issues and organizational structures in industry today.

a. Describe the leadership issues facing the construction industry.
b. Explain how gender and cultural issues affect the construction industry.
c. Explain the organization of construction businesses and the need for policies and procedures.

Trade Terms

Cloud-based applications: Mobile and desktop digital programs that can connect to files and data located in distributed storage locations on the Internet ("the Cloud"). Such applications make it possible for authorized users to create, edit, and distribute content from any location where Internet access is possible.

Craft professionals: Workers who are properly trained and work in a particular construction trade or craft.

Crew leader: The immediate supervisor of a crew or team of craft professionals and other assigned persons.

Demographics: Social characteristics and other factors, such as language, economics, education, culture, and age, that define a statistical group of individuals. An individual can be a member of more than one demographic.

Job description: A description of the scope and responsibilities of a worker's job so that the individual and others understand what the job entails.

Letter of instruction (LOI): A written communication from a supervisor to a subordinate that informs the latter of some inadequacy in the individual's performance, and provides a list of actions that the individual must satisfactorily complete to remediate the problem. Usually a key step in a series of disciplinary actions.

Local area networks (LAN): Communication networks that link computers, printers, and servers within a small, defined location, such as a building or office, via hard-wired or wireless connections.

Organizational chart: A diagram that shows how the various management and operational responsibilities relate to each other within an organization. Named positions appear in ranked levels, top to bottom, from those functional units with the most and broadest authority to those with the least, with lines connecting positions indicating chains of authority and other relationships.

Paraphrase: Rewording a written or verbal statement in one's own words in such a way that the intent of the original statement is retained.

Pragmatic: Sensible, practical, and realistic.

Sexual harassment: Any unwelcome verbal or nonverbal form of communication or action construed by an individual to be of a sexual or gender-related nature.

Smartphone: The most common type of cellular telephone that combines many other digital functions within a single mobile device. Most important of these, besides the wireless telephone, are high-resolution still and video cameras, text and email messaging, and GPS-enabled features.

Synergy: Any type of cooperation between organizations, individuals, or other entities where the combined effect is greater than the sum of the individual efforts.

Wide area networks (WAN): Dedicated communication networks of computers and related hardware that serve a given geographic area, such as a work site, campus, city, or a larger but distinct area. Connectivity is by wired and wireless means, and may use the Internet as well.

Wi-Fi: The technology allowing communications via radio signals over a LAN or WAN equipped with a wireless access point or the Internet. (Wi-Fi stands for wireless fidelity.) Many types of mobile, portable, and desktop devices can communicate via Wi-Fi connections.

Today's managers, supervisors, and lower-level managers face challenges different from those of previous generations. To be a crew leader today, it is essential to be well prepared. Crew leaders must understand how to use various types of new technology. In addition, they must have the knowledge and skills needed to manage, train, and communicate with a culturally-diverse workforce whose attitudes toward work and job expectations may differ from those of earlier generations and cultures.

A summary of the changes in the workforce, in the work environment, and of industry needs include the following:

- A shrinking workforce
- The growth of construction, communication, and scheduling technology
- Changes in the attitudes and values of **craft professionals**
- The rapidly-changing **demographics** of gender, gender-identity, and foreign-born workers
- Increased emphasis on workplace safety and health
- Greater need for more education and training

1.1.0 Leadership Issues and Training Strategies

Effective craft training programs are necessary if the industry is to meet the forecasted worker demands. Many skilled, knowledgeable craft professionals, crew leaders, and managers have reached retirement age. In 2015, the generation of workers called *"Baby Boomers,"* who were born between 1946 and 1964, represented 50 percent of the workforce. Their departure creates a large demand for craftworkers across the industry. The US Department of Labor (DOL) concludes that the best way for industry to reduce shortages of skilled workers is to create more education and training opportunities. The DOL suggests that companies and community groups form partnerships and create apprenticeship programs. Such programs could provide younger workers, including women and minorities, with the opportunity to develop job skills by giving them hands-on experience.

When training workers, it is important to understand that people learn in different ways. Some people learn by doing, some people learn by watching or reading, and others need step-by-step instructions as they are shown the process. Most people learn best through a combination of styles. While you may have the tendency to teach in the style that you learn best, you must always stay in tune with what kind of learner you are teaching. Have you ever tried to teach somebody and failed, and then another person successfully teaches the same thing in a different way? A person who acts as a mentor or trainer needs to be able to determine what kind of learner they are addressing, and teach according to those needs.

"The mediocre teacher tells.
The good teacher explains.
The superior teacher demonstrates.
The great teacher inspires."
—William Arthur Ward (1921)

The need for training isn't limited to craft professionals. There must be supervisory training to ensure there are qualified leaders in the industry to supervise the craft professionals.

1.1.1 Motivation

As a supervisor or crew leader, it is important to understand what motivates your crew. Money is often considered a good motivator, but it is sometimes only a temporary solution. Once a person has reached a level of financial security, other factors come into play. Studies show that environment and conditions motivate many people. For those people, a great workplace may mean more to them than better pay.

If you give someone a raise, they tend to work harder for a period of time. Then the satisfaction wanes and they may want another raise. A sense of accomplishment is what motivates most people. That is why setting and working toward recognizable goals tends to make employees more productive. A person with a feeling of involvement or a sense of achievement is likely to be better motivated and help to motivate others.

1.1.2 Understanding Workers

Many older workers grew up in an environment where they learned to work hard with the same employer until retirement. They expected to stay with a company for a long time, and the structure of the companies created a family-type environment.

Times have changed. Younger workers have grown up in a highly mobile society and expect frequent rewards and rapid advancement. Some might perceive this generation of workers as lazy, narcissistic, and unmotivated, but in reality, they simply have a different perspective on life, work, and priorities. For such workers, it may be better to give them small projects or break up large projects into smaller pieces so that they are rewarded more often by successfully achieving short-term goals. The following strategies are important for keeping young workers motivated and engaged:

- *Goal setting* – Set short-term and long-term goals, including tasks to be done and expected time frames. Things can happen to change, delay, or upset the short-term goals. This is one reason to set long-term goals as well. Don't set workers up for failure, as this leads to frustration, and frustration can lead to reduced productivity.
- *Feedback* – Timely feedback is important. For example, telling someone they did a good job last year, or criticizing them for a job they did a month ago, is meaningless. Simple recognition isn't always enough. Some type of reward should accompany positive feedback, even if it is simply recognizing the employee in a public way. Constructive criticism or reprimands should always be given in private. You can also provide some positive action, such as one-on-one training, to correct a problem.

1.1.3 Craft Training

Craft training is often informal, taking place on the job site, outside of a traditional training classroom. According to the American Society for Training and Development (ASTD), a qualified co-worker or a supervisor conducts craft training generally through on-the-job instruction. The Society of Human Resources Management (SHRM) offers the following tips to supervisors in charge of training their employees:

- *Help crew members establish career goals* – Once career goals are established, you can readily identify the training required to meet the goals.
- *Determine what kind of training to give* – Training can be on the job under the supervision of a co-worker. It can be one-on-one with the supervisor. It can involve cross training to teach a new trade or skill, or it can involve delegating new or additional responsibilities.
- *Determine the trainee's preferred method of learning* – Some people learn best by watching, others from verbal instructions, and others by doing. These are categorized as visual learners, auditory learners, and tactile learners, respectively. When training more than one person at a time, try to use a mix of all three methods.

Communication is a critical component of training employees. SHRM advises that supervisors do the following when training their employees:

- Explain the task, why it is important, and how to do it. Confirm that the worker trainees understand these three areas by asking questions. Allow them to ask questions as well.
- Demonstrate the task. Break the task down into manageable parts and cover one part at a time.
- Ask your trainees to do the task while you observe them. Try not to interrupt them while they are doing the task unless they are doing something that is unsafe or potentially harmful.
- Give the trainees feedback. Be specific about what they did and mention any areas where they need to improve.

1.1.4 Supervisory Training

Because of the need for skilled craft professionals and qualified supervisory personnel, some companies offer training to their employees through in-house classes, or by subsidizing outside training programs. However, for a variety of reasons, many contractors don't offer training at all. Some common reasons include the following:

- Lack of money to train
- Lack of time to train
- Lack of knowledge about the benefits of training programs
- High rate of employee turnover
- Workforce is too small
- Past training involvement was ineffective
- The company hires only trained workers
- Lack of interest from workers

For craft professionals to move up into supervisory and managerial positions, they must continue their education and training. Those who are willing to acquire and develop new skills have the best chance of finding stable employment. This makes it critical for them to take advantage of training opportunities, and for companies to incorporate training into their business culture.

If your company has recognized the need for training, your participation in a leadership training program such as this will begin to fill the gap between craft and supervisory training.

1.1.5 Impact of Technology

Many industries, including the construction industry, have embraced technology to remain competitive. Benefits of technology include increased productivity and speed, improved quality of documents, greater access to common data, and better financial controls and communication. As technology becomes a greater part of supervision, crew leaders must be able to use it properly.

Cellphones (in particular the smartphone) have made it easy to keep in touch through numerous forms of media. As of 2016, 95 percent of all Americans own a cellphone of some kind

Did You Know?

Learning Strategies

There is a lot more to training than simply telling a group of people what you know about a subject or talking your way through a series of slides. Studies have shown that the information and understanding that learners retain depends on how they receive the information and use it. These learning processes are as follows, listed from least effective to most effective:

1. Reading about a process
2. Hearing a description of a process
3. Observing a process
4. Observing and hearing a description of a process
5. Observing, hearing, and responding to instruction
6. Observing, hearing, and doing a process

Although some companies may attempt to assign percentage effectiveness to these instructional approaches, the results are very arbitrary and inconclusive. It is more important to understand which are the most effective methods for teaching and training.

and 77 percent own a smartphone, according to the Pew Research Center. They are particularly useful communication tools for contractors or crew leaders who are on a job site, away from their offices, or constantly on the go. Workers use smartphones at any time for phone calls, emails, text messages, and voicemail, as well as to share photos and videos. The hundreds of thousands of mobile computing apps available allow smartphones to perform numerous other functions on the go. However, the number of accidents due to inattention while focusing on a smartphone or other mobile device is also rapidly rising. Always check the company's policy regarding cell phone use on the job.

As a crew leader, you should be aware that smartphones and tablets (*Figure 1*) allow supervisors to plan their calendars, schedule meetings, manage projects, and access their company email from remote locations. These devices are far more powerful than the computers that took us to the moon, and they can hold years of information from various projects. **Cloud-based applications** now permit remote access to and updating of files, plans, and data from any place with wireless connectivity. In fact, it is becoming common for work sites to set up **local area networks (LAN)**, **wide area networks (WAN)**, and dedicated **Wi-Fi** services to support mobile communications among workers on the job.

In all forms of electronic communication (verbal, written, and visual), it is important to keep messages brief, factual, and legal. Text-based communications can be easily misunderstood because there are no visual or auditory cues to indicate the sender's intent. In other words, it is more difficult to tell if someone is just joking via email because you can't see the sender's expression or hear their tone of voice.

1.2.0 Gender and Cultural Issues

In the past several years, the construction industry in the United States has experienced a shift in worker expectations and diversity. These two issues are converging at a rapid pace.

The generation of learners is also a factor in the learning process and in the workplace. The various generations include Baby Boomers, Generation X, Millennials (Generation Y), and Generation Z. The ranges of birth years for these generations are as follows (note that there may be

Figure 1 Digital tablets are a convenient management and scheduling tool for supervisors.

some overlap, and opinions on the generational names as well as the range of years for each generation may vary):

- *Baby Boomers* – born 1946 through 1964
- *Generation X* – born 1965 through 1979
- *Millennials (Generation Y)* – born 1980 through 2000
- *Generation Z* – born 2000 through 2015

Each generation has been studied to some degree to determine their different interpretations of and approaches to learning and training. Family norms, religious values and morals, educational methods, music, movies, politics, and global events are all influential factors that define a generation. Remember that individuals within a generation can vary widely in their expectations and behaviors.

This trend, combined with industry diversity initiatives, has created a climate in which companies recognize the need to embrace a diverse workforce that crosses generational, gender, and ethnic boundaries (*Figure 2*). To do this effectively, they are using their own resources, as well as relying on associations with the government and trade organizations.

All current research indicates that industry will be more dependent on the critical skills of a diverse workforce—but a workforce that must be both culturally and ethnically fused. Across the United States, construction and other industries are aggressively seeking to bring new workers into their ranks, including women, racial, and ethnic minorities. Social and political issues are no longer the main factors driving workplace diversity, but by consumers and citizens who need more hospitals, malls, bridges, power plants, refineries, and many other commercial and residential structures. The construction industry needs more workers.

Figure 2 The modern workforce is diverse.

There are some potential issues relating to a diverse workforce that may be encountered on the job site. These issues include different communication styles of men and women, language barriers associated with cultural differences, **sexual harassment**, and gender or racial discrimination.

1.2.1 Communication Styles of Men and Women

As more and more women enter construction workforce, it becomes increasingly important to break down communication barriers between men and women and to understand differences in behaviors so that men and women can work together more effectively. The Jamestown, Area Labor Management Committee (JALMC) in New York offers the following explanations and tips (*emphasized text* is from *Exchange and Deception: A Feminist Perspective*, ed. by Caroline Gerschlager, Monika Mokre, 2002. Springer Science & Business Media):

- *Women tend to ask more questions than men do* – Men are more likely to proceed with a job and figure it out as they go along, while women are more likely to ask questions first.
- *Men tend to offer solutions before empathy; women tend to do the opposite* – Both men and women should say what they want up front, whether it's the solution to a problem, or simply a sympathetic ear. That way, both genders will feel understood and supported.
- *Women are more likely to ask for help when they need it* – Women are generally more **pragmatic** when it comes to completing a task. If they need help, they will ask for it. Men are more likely to attempt to complete a task by themselves, even when they need assistance.
- *Men tend to communicate more competitively, and women tend to communicate more cooperatively* – Both parties need to hear one another out without interruption.

This doesn't mean that any one method is better or worse than the other; it simply means that men and women inherently use different approaches to achieve the same result. It's either genetic or cultural, but awareness can overcome these tendencies.

1.2.2 Language Barriers

Language barriers are a real workplace challenge for crew leaders. Millions of American workers speak languages other than English. Bilingual job sites are increasingly common. As the makeup of the immigrant population continues to change with the influx of refugees from around the world, the number of non-English speakers in the American workforce is rising and diversifying dramatically.

In addition, there are many different dialects of the English language in America, which can present some communication challenges. For example, some workers may speak a dialect commonly called *Ebonics* (referred to by linguists as *African-American Vernacular English*). The dialect called *Rural White Southern English* can also be difficult to understand by those not born in the American South. Many find the American New England accent and dialect difficult to comprehend. Some sources list as many as 16 distinct American dialects.

Companies have the following options to overcome the language challenge, mainly for foreign or English-as-a-second-language (ESL) workers:

- Offer English classes either at the work site or through school districts and community colleges.
- Offer incentives for workers to learn English.

Communication with ESL workers, and varying dialects in general, becomes even more critical as the workforce grows more diverse. The following tips will help when communicating across language barriers:

- Be patient. Give workers time to process the information in a way they can comprehend.
- Avoid humor. Humor is easily misunderstood. The worker may misinterpret what you say as a joke at the worker's expense.
- Don't assume workers are unintelligent simply because they don't understand what you are saying. Explaining something in multiple ways, and having trainees **paraphrase** what they heard, is an excellent way to ensure mutual understanding and prevent miscommunication.
- If a worker is not fluent in English, ask the worker to demonstrate his or her understanding through other means.
- Speak slowly and clearly, and avoid the tendency to raise your voice.
- Use face-to-face communication whenever possible. Over-the-phone communication is often more difficult when a language barrier is involved.
- Use pictures or sketches to get your point across.

1.2.3 Cultural Differences

As workers from a multitude of backgrounds and cultures work together, there are bound to be differences and conflicts in the workplace. To overcome cultural conflicts, the SHRM suggests the following approach to resolving cultural conflicts between individuals:

- *Define the problem from both points of view* – How does each person involved view the conflict? What does each person think is wrong? This involves moving beyond traditional thought processes to consider alternate ways of thinking.
- *Uncover cultural interpretations* – What assumptions may the parties involved be making based on cultural programming? This is particularly true for certain gestures, symbols, and words that mean different things in different cultures. By doing this, the supervisor may realize what motivated an employee to act in a specific way.
- *Create cultural* **synergy** – Devise a solution that works for both parties involved. The purpose is to recognize and respect other's cultural values, and work out mutually acceptable alternatives.

1.2.4 Sexual Harassment

In today's business world, men and women are working side-by-side in careers of all kinds, creating increased opportunity for sexual harassment to occur. Sexual harassment can be defined as any unwelcome behavior that makes someone feel uncomfortable in the workplace by focusing attention on their gender or gender identity. Activities that might qualify as sexual harassment include, but are not limited to, the following:

- Telling an offensive, sexually-oriented joke
- Displaying a poster of a man or woman in a revealing swimsuit
- Wearing a patch or article of clothing blatantly promoting or degrading a specific gender
- Making verbal or physical advances (*Figure 3*)
- Speaking abusively about a specific gender

Historically, many have thought of sexual harassment as an act perpetrated by men against women, especially those in subordinate positions. However, the nature of sexual harassment cases over the years have shown that the perpetrator can be of any gender.

Figure 3 Any unwelcome physical contact can be considered harassment.

Sexual harassment can occur in various circumstances, including the following:

- The victim and harasser may be either male or female. The victim does not have to be of the opposite gender.
- The harasser can be the victim's supervisor, an agent of the employer, a supervisor in another area, a co-worker, a subordinate, or a nonemployee.
- Legally, the victim doesn't have to be the person harassed, but could be anyone offended by the relevant conduct.
- Unlawful sexual harassment may occur without economic injury to or the discharge of the victim.
- The harasser's conduct must be unwelcome.

The Equal Employment Opportunity Commission (EEOC) enforces sexual harassment laws within industries. When investigating allegations of sexual harassment, the EEOC looks at the whole record, including the circumstances and the context in which the alleged incidents occurred. A decision on the allegations is made from the facts on a case-by-case basis. Supervision can hold the crew leader responsible who is aware of sexual harassment and does nothing to stop it. The crew leader therefore should not only take action to stop sexual harassment, but should serve as a good example for the rest of the crew.

Prevention is the best tool to eliminate sexual harassment in the workplace. The EEOC encourages employers to take steps to prevent sexual harassment from occurring. Employers should clearly communicate to employees that they and their company won't tolerate sexual harassment. They do so by developing a policy on sexual harassment, establishing an effective complaint or grievance process, and taking immediate and appropriate action when an employee complains.

Controversial or sexually-related remarks, jokes, and swearing are not only offensive to co-workers, but also tarnish a worker's character. Crew leaders need to emphasize that abrasive or crude behavior may affect opportunities for advancement. If disciplinary action becomes necessary, written company policies should spell out the type of actions warranted by such behavior. A typical approach is a three-step process in which the perpetrator is first given a verbal reprimand. In the event of further violations, a written reprimand, often including a **letter of instruction (LOI)** and a warning are given. Dismissal typically accompanies any subsequent violations.

1.2.5 Gender and Minority Discrimination

Employers are giving more attention to fair recruitment, equal pay for equal work, and promotions for women and minorities in the workplace. Consequently, many businesses are analyzing their practices for equity, including the treatment of employees, the organization's hiring and promotional practices, and compensation.

The construction industry, which was once male-dominated, is moving away from this image and actively recruiting and training women, younger workers, people from other cultures, and disabled workers. This means that organizations hire the best person for the job, without regard for race, sex, religion, age, etc.

Did You Know?

Respectful Workplace Training

Some companies employ a tool called *sensitivity training* in cases where individuals or groups have trouble adapting to a multi-cultural, multi-gender workforce. Sensitivity training is a psychological technique using group discussion, role playing, and other methods to allow participants to develop an awareness of themselves and how they interact with others. Critics of earlier forms of sensitivity training noted that the methods used were not very different from brainwashing. Modern forms of cultural-awareness and gender-issue training are more ethical, and go by the term *respectful workplace training.*

To prevent discrimination cases, employers must have valid job-related criteria for hiring, compensation, and promotion. They must apply these measures consistently for every applicant interview, employee performance appraisal, and hiring or promotion decision. Therefore, employers must train all workers responsible for recruitment, selection, supervision of employees, and evaluating job performance on how to use the job-related criteria legally and effectively.

1.3.0 Business Organization

An organization is the relationship among the people within the company or project. The crew leader needs to be aware of two types of organizations: formal organizations and informal organizations.

A formal organization exists when the people within a work group led by someone direct their activities toward achieving a common goal. An example of a formal organization is a work crew consisting of four carpenters and two laborers led by a crew leader, all working together to accomplish a project.

An **organizational chart** typically illustrates a formal organization. It outlines all the positions that make up an organization and shows how those positions are related. Some organizational charts even identify the person within each position and the person to whom that person reports, as well as the people that the person supervises. *Figure 4* and *Figure 5* show examples of what an organization chart might look like for a construction company and an industrial company. Note that each organizational position represents an opportunity for advancement in the construction industry that a crew leader can eventually achieve.

An informal organization allows for communication among its members so they can perform as a group. It also establishes patterns of behavior that help them to work as a group, such as agreeing to use a specific training or certification program.

An example of an informal organization is a trade association such as Associated Builders and Contractors (ABC), Associated General Contractors (AGC), and the National Association of Women in Construction (NAWIC). Those, along with the thousands of other trade associations in the United States, provide forums in which members with common concerns can share information, work on issues, and develop standards for their industry.

Both formal and informal organizations establish the foundation for how communication flows. The formal structure is the means used to delegate authority and responsibility and to exchange information. The informal structure provides means for the exchange of information.

Members in an organization perform best when they can do the following:

- Know the job and how it will be done
- Communicate effectively with others in the group
- Understand their role in the organization
- Recognize who has the authority and responsibility

"Leadership is not about titles, positions, or flow charts. It is about one life influencing another."
—John C. Maxwell

1.3.1 Division of Responsibility

The conduct of a business involves certain functions. In a small organization, one or two people may share responsibilities. However, in a larger organization with many different and complex activities, distinct activity groups may lie under the responsibility of separate department managers. In either case, the following major department functions exist in most companies:

- *Executive* – This office represents top management. It is responsible for the success of the company through short-range and long-range planning.
- *Human Resources (HR)* – This office is responsible for recruiting and screening prospective employees, managing employee benefits programs, advising management on pay and benefits, and developing and enforcing procedures related to hiring practices.
- *Accounting* – This office is responsible for all recordkeeping and financial transactions, including payroll, taxes, insurance, and audits.
- *Contract Administration* – This office prepares and executes contractual documents with owners, subcontractors, and suppliers.
- *Purchasing* – This office obtains material prices and then issues purchase orders. The purchasing office also obtains rental and leasing rates on equipment and tools.
- *Estimating* – This office is responsible for recording the quantity of material on the jobs, the takeoff, pricing labor and material, analyzing subcontractor bids, and bidding on projects.
- *Operations* – This office plans, controls, and supervises all project-related activities.

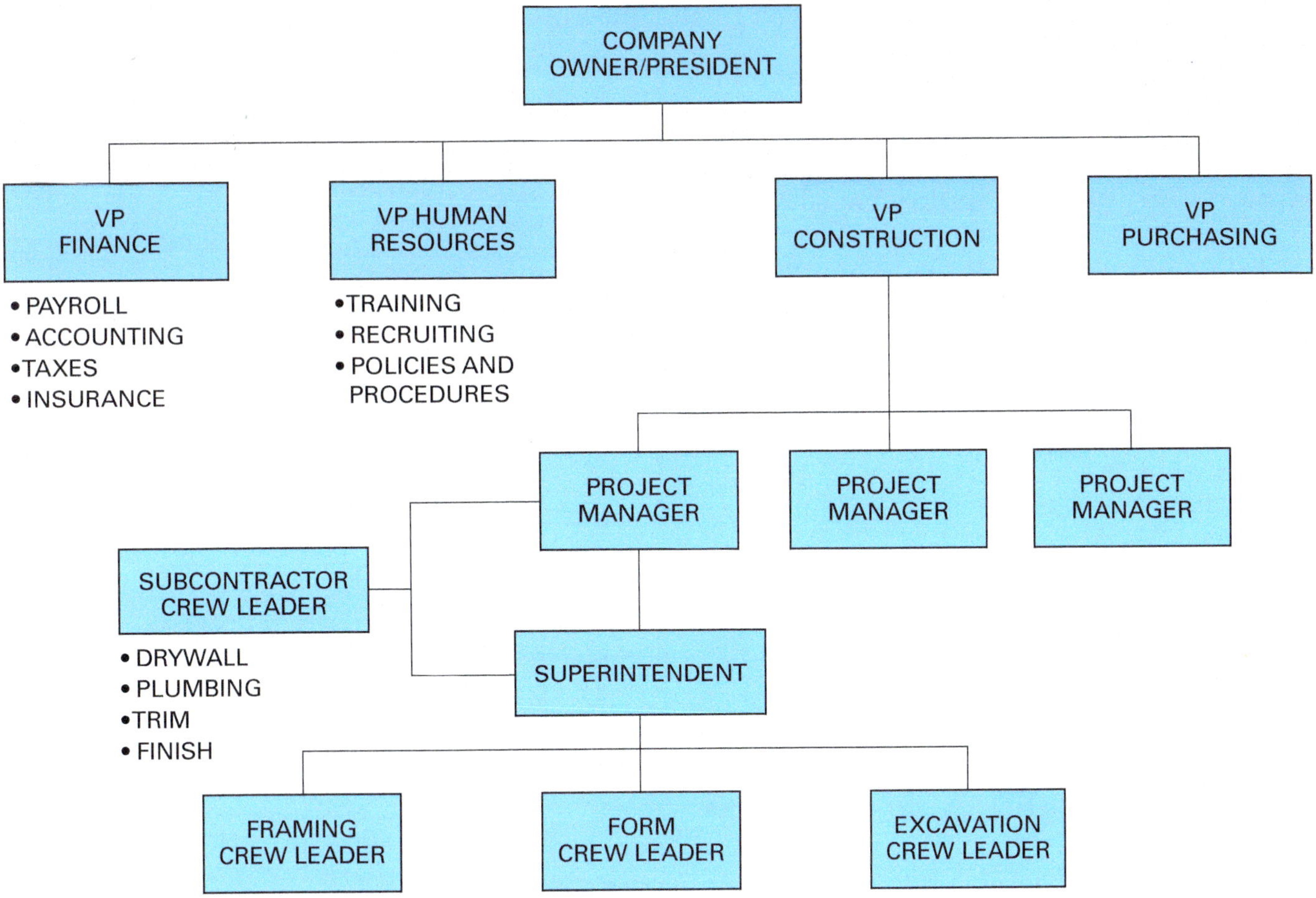

Figure 4 Sample organization chart for a construction company.

Other divisions of responsibility a company may create involve architectural and engineering design functions. These divisions usually become separate departments.

1.3.2 Responsibility, Authority, and Accountability

As an organization grows, the manager must ask others to perform many duties so that the manager can concentrate on management tasks. Managers typically assign (delegate) activities to their subordinates. When delegating activities, the crew leader assigns others the responsibility to perform the designated tasks.

Responsibility is the obligation to perform the duties. Along with responsibility comes authority. *Authority* is the right to act or make decisions in carrying out an assignment. The type and amount of authority a supervisor or worker has depends on the employee's company. There must be a balance between the authority and responsibility workers have so that they can carry out their assigned tasks. In addition, the crew leader must delegate sufficient authority to make a worker accountable to the crew leader for the results.

Accountability is holding an employee responsible for completing the assigned activities. Even though the crew leader may delegate authority and responsibility to crew members, the overall responsibility—the accountability—for the tasks assigned to the crew always rests with the crew leader.

1.3.3 Job Descriptions

Many companies furnish each employee with a written **job description** that explains the job in detail. Job descriptions set a standard for the employee. They make judging performance easier, clarify the tasks each person should handle, and simplify the training of new employees.

Each new employee should understand all the duties and responsibilities of the job after reviewing the job description. Thus, the time is shorter for the employee to make the transition from being a new and uninformed employee to a more experienced member of a crew.

A job description need not be long, but it should be detailed enough to ensure there is no misunderstanding of the duties and responsibilities of the position. The job description should

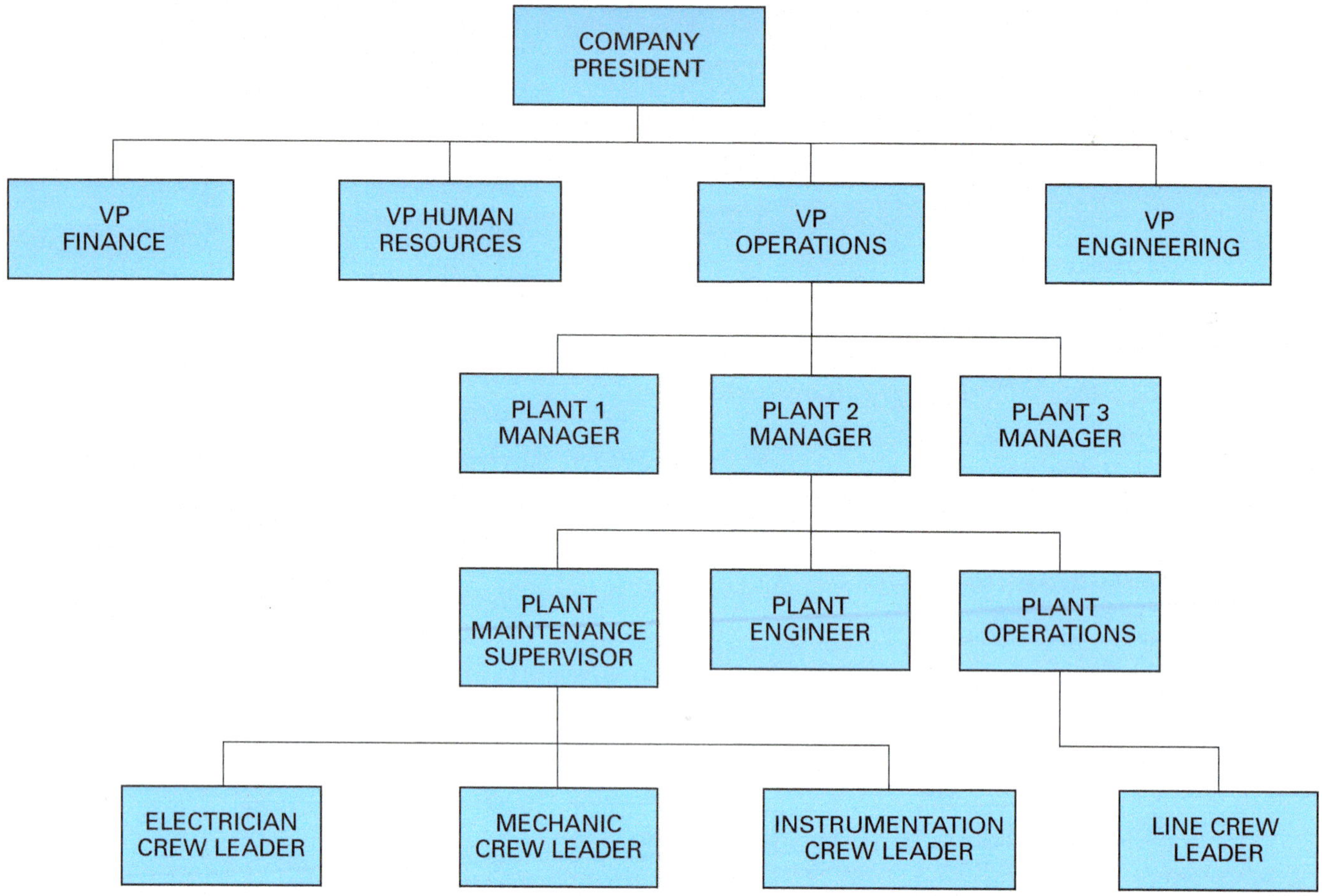

Figure 5 Sample organization chart for an industrial company.

contain all the information necessary to evaluate the employee's performance and hold the employee accountable.

A job description should contain, at minimum, the following:

- Job title
- General description of the position
- Minimum qualifications for the job
- Specific duties and responsibilities
- The supervisor to whom the position reports
- Other requirements, such as qualifications, certifications, and licenses

Figure 6 is an example of a job description.

1.3.4 Policies and Procedures

Most companies have formal policies and procedures established to help crew leaders carry out their duties. A policy is a general statement establishing guidelines for a specific activity. Examples include policies on vacations, breaks, workplace safety, and checking out tools. Procedures are formal instructions to carry out and meet policies. For example, a procedure written to implement a policy on workplace safety would include guidelines expected of all employees for reporting accidents and to follow general safety practices.

A crew leader must be familiar with the company policies and procedures, especially regarding safety practices. When OSHA inspectors visit a jobsite, they often question employees and crew leaders about the company policies related to safety. If they are investigating an accident, they will want to verify that the responsible crew leader knew the applicable company policy and followed it.

Position:
Crew Leader

General Summary:
First line of supervision on a construction crew installing concrete formwork.

Reports To:
Job Superintendent

Physical and Mental Responsibilities:
- Ability to stand for long periods
- Ability to solve basic math and geometry problems

Duties and Responsibilities:
- Oversee crew
- Provide instruction and training in construction tasks as needed
- Make sure proper materials and tools are on the site to accomplish tasks
- Keep project on schedule
- Enforce safety policies and procedures

Knowledge, Skills, and Experience Required:
- Extensive travel throughout the Eastern United States, home base in Atlanta
- Ability to operate a backhoe and trencher
- Valid commercial driver's license with no DUI violations
- Ability to work under deadlines with the knowledge and ability to foresee problem areas and develop a plan of action to solve the situation

Figure 6 A sample job description.

Did You Know?

The ADA

The Americans with Disabilities Act (ADA) is a law that prohibits employers from discriminating against disabled persons in their hiring practices. While the ADA doesn't require written job descriptions, investigators use existing job descriptions as evidence in determining whether discrimination occurred. This means that job descriptions must specifically define the job duties and the abilities needed to perform them. In making a determination of discrimination under the ADA, investigators give consideration to whether a reasonable accommodation could have qualified the disabled person to fill a job.

Additional Resources

Construction Workforce Development Professional, NCCER. 2016. New York, NY: Pearson Education, Inc.

Mentoring for Craft Professionals, NCCER. 2016. New York, NY: Pearson Education, Inc.

Generational Cohorts and their Attitudes Toward Work Related Issues in Central Kentucky, Frank Fletcher, et al. 2009. Midway College, Midway, KY. **www.kentucky.com**

The Young Person's Guide to Wisdom, Power, and Life Success: Making Smart Choices. Brian Gahran, PhD. 2014. San Diego, CA: Young Persons Press. **www.WPGBlog.com**

The following websites offer resources for products and training:

Aging Workforce News, **www.agingworkforcenews.com**

American Society for Training and Development (ASTD), **www.astd.org**

Equal Employment Opportunity Commission (EEOC), **www.eeoc.gov**

National Association of Women in Construction (NAWIC), **www.nawic.org**

Society for Human Resources Management (SHRM), **www.shrm.org**

United States Census Bureau, **www.census.gov**

United States Department of Labor, **www.dol.gov**

Wi-Fi® is a registered trademark of the Wi-Fi Alliance, **www.wi-fi.org**

1.0.0 Section Review

1. According to the US Department of Labor, the best way for the construction industry to reduce skilled-worker shortages is to _____.
   a. create training opportunities
   b. avoid discrimination lawsuits
   c. update the skills of older workers who are retiring at a later age than they previously did
   d. implement better policies and procedures

2. Which of the following is *not* helpful when dealing with language diversity in the workplace?
   a. Speaking slowly and clearly in an even tone
   b. Using humor when communicating
   c. Using sketches or diagrams to explain what needs to be done.
   d. Being patient and giving bilingual workers time to process your instructions.

3. Members tend to function best within an organization when they _____.
   a. are allowed to select their own uniform for each project
   b. understand their role within the organization
   c. don't disagree with the statements of other workers or supervisors
   d. are able to work without supervision

Section Two

2.0.0 Leadership Skills

Objective

Explain how to incorporate leadership skills into work habits, including communications, motivation, team-building, problem-solving, and decision-making skills.

a. Describe the role of a leader on a construction crew.
b. Explain the importance of written and oral communication skills.
c. Describe methods for motivating team members.
d. Explain the importance of teamwork to a construction project.
e. Identify effective problem-solving and decision-making methods.

Trade Terms

Autonomy: The condition of having complete control over one's actions, and being free from the control of another. To be independent.

Bias: A preconceived inclination against or in favor of something.

Ethics: The moral principles that guides an individual's or organization's actions when dealing with others. Also refers to the study of moral principles.

Infer: To reach a conclusion using a method of reasoning that starts with an assumption and considers a set of logically-related events, conditions, or statements.

Legend: In maps, plans, and diagrams, an explanatory table defining all symbolic information contained in the document.

Lockout/tagout (LOTO): A system of safety procedures for securing electrical and mechanical equipment during repairs or construction, consisting of warning tags and physical locking or restraining devices applied to controls to prevent the accidental or purposeful operation of the equipment, endangering workers, equipment, or facilities.

Proactive: To anticipate and take action in the present to deal with potential future events or outcomes based on what one knows about current events or conditions.

Project manager: In construction, the individual who has overall responsibility for one or more construction projects; also called the general superintendent.

Superintendent: In construction, the individual who is the on-site supervisor in charge of a given construction project.

Textspeak: A form of written language characteristic of text messaging on mobile devices and text-based social media, which usually consists of acronyms, abbreviations, and minimal punctuation.

It is important to define some of the supervisory positions discussed throughout this module. You are already familiar with the roles of the craft professional and crew leader. A superintendent is essentially an on-site supervisor who is responsible for one or more crew leaders or front-line supervisors. A project manager or general superintendent may be responsible for managing one or more projects. This training will concentrate primarily on the supervisory role of the crew leader.

Craftworkers and crew leaders differ in that the crew leader manages the activities that the craft professionals perform. To manage a crew of craft professionals, a crew leader must have first-hand knowledge and experience in their activities. Additionally, the crew leader must be able to act directly in organizing and directing the activities of the various crew members.

This section explains the importance of developing leadership skills as a new crew leader. It will cover effective ways to communicate with coworkers and employees at all levels, build teams, motivate crew members, make decisions, and resolve problems.

Crew leaders are generally promoted up from a work crew. A worker's ability to accomplish tasks, get along with others, meet schedules, and stay within the budget have a significant influence on the selection process. The crew leader must lead the team to work safely and provide a quality product.

Making the transition from crew member to a crew leader can be difficult, especially when the new position involves overseeing a group of former peers. For example, some of the crew may try to take advantage of their friendship by seeking special favors. They may also want to be privy to supervisory information that is normally closely held. When you become a crew leader, you are no longer responsible for your work alone. Crew leaders are accountable for the work of an entire crew of people with varying skill levels, personalities, work styles, and cultural and educational backgrounds.

Crew leaders must learn to put personal relationships aside and work for the common goals of the entire crew. The crew leader can overcome these problems by working with the crew to set mutual performance goals and by freely communicating with them within permitted limits. Use their knowledge and strengths along with your own so that they feel like they are key players on the team.

As employees move from being a craftworker into the role of a crew leader and above, they will begin to spend more hours supervising the work of others (supervisory work) than practicing their own craft skills (technical work). *Figure 7* illustrates the relative amounts of time craft professionals, crew leaders, superintendents, and project managers spend on technical and supervisory work as their management responsibilities increase.

There are many ways to define a leader. One simple definition of a leader is a person who influences other people to achieve a goal. Some people may have innate leadership qualities that developed during their upbringing, or they may have worked to develop the traits that motivate others to follow and perform. Research shows that people who possess such talents are likely to succeed as leaders.

"Leadership is the art of getting someone else to do something you want done because he wants to do it."
—Dwight D. Eisenhower

2.1.0 The Qualities and Role of a Leader

Leadership traits are similar to the skills that a crew leader needs to be effective. Although the characteristics of leadership are many, there are some definite commonalities among effective leaders.

First and foremost, effective leaders lead by example. They work and live by the standards that they establish for their crew members or followers, making sure they set a positive example.

Effective leaders also tend to have a high level of drive and determination, as well as a persistent attitude. When faced with obstacles, effective leaders don't get discouraged. Instead, they identify the potential problems, make plans to overcome them, and work toward achieving the intended goal. In the event of failure, effective leaders learn from their mistakes and apply that knowledge to future situations. They also learn from their successes.

Effective leaders are typically good communicators who clearly lay out the goals of a project to their crew members. Accomplishing this may require that the leader overcome issues such as language barriers, gender bias, or differences in personalities to ensure that each member of the crew understands the established goals of the project.

Effective leaders can motivate their crew members to work to their full potential and become useful members of the team. Crew leaders try to develop crew-member skills and encourage them to improve and learn so they can contribute more to the team effort. Effective leaders strive for excellence from themselves and their team, so they work hard to provide the skills and leadership necessary to do so.

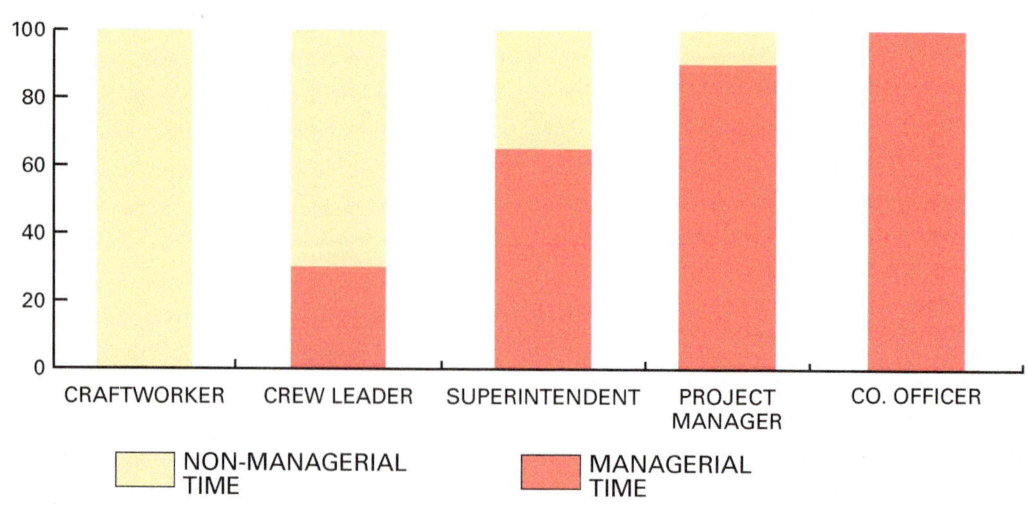

Figure 7 Percentages of technical and supervisory work by role.

In addition, effective leaders must possess organizational skills. They know what needs to be accomplished, and they use their resources to make it happen. Because they can't do it alone, leaders require the help of their team members to share in the workload. Effective leaders delegate work to their crew members, and they implement company policies and procedures to complete the work safely, effectively, and efficiently.

Finally, effective leaders have the experience, authority, and self-confidence that allows them to make decisions and solve problems. To accomplish their goals, leaders must be able to calculate risks, take in and interpret information, assess courses of action, make decisions, and assume the responsibility for those decisions.

2.1.1 Functions of a Leader

The functions of a leader will vary with the environment, the group of individuals they lead, and the tasks they must perform. However, there are certain functions common to all situations that the leader must fulfill. Some of the major functions are as follows:

- Accept responsibility for the successes and failures of the group's performance
- Be sensitive to the differences of a diverse workforce
- Ensure that all group members understand and abide by company policies and procedures
- Give group members the confidence to make decisions and take responsibility for their work
- Maintain a cohesive group by resolving tensions and differences among its members and between the group and those outside the group
- Organize, plan, staff, direct, and control work
- Represent the group

2.1.2 Leadership Traits

There are many traits and skills that help create an effective leader. Some of these include the following:

- Ability to advocate an idea
- Ability to motivate
- Ability to plan and organize
- Ability to teach others
- Enthusiasm
- Fairness
- Good communication skills
- Initiative
- Loyalty to their company and crew
- Willingness to learn from others

2.1.3 Expected Leadership Behavior

Followers have expectations of their leaders. They look to their leaders to do the following:

- Abide by company policies and procedures
- Be a loyal member of the team
- Communicate effectively
- Have the necessary technical knowledge
- Lead by example
- Make decisions and assume responsibility
- Plan and organize the work
- Suggest and direct
- Trust the team members

2.1.4 Leadership Styles

Through history, there have been many terms used to describe different leadership styles and the amount of **autonomy** they permit. The amount of crew autonomy directly relates to the leadership style used by the crew leader. *Figure 8* illustrates the linear relationship between leader authority and worker autonomy.

There are three main styles of leadership. At one extreme is the *controller style* of leadership, where the crew leader makes all of the decisions independently, without seeking the opinions of crew members. (Very little autonomy exists for workers under this type of leader.) At the other extreme is the *advisor style*, where the crew leader empowers the employees to make decisions (high autonomy). In between these extremes is the *directive style*, where the crew leader seeks crew member opinions and makes the appropriate decisions based on their input.

The following are some characteristics of each of the three leadership styles:

Controller style (high authority, low autonomy):

- Expect crew members to work without questioning procedures
- Seldom seek advice from crew members
- Insist on solving problems alone
- Seldom permit crew members to assist each other
- Praise and criticize on a personal basis
- Have no sincere interest in creatively improving methods of operation or production

Directive style (equal authority and autonomy):

- Discuss problems with their crew members
- Listen to suggestions from crew members
- Explain and instruct
- Give crew members a feeling of accomplishment by commending them when they do a job well

- Are friendly and available to discuss personal and job-related problems

Advisor style (low authority, high autonomy):

- Believe no supervision is best
- Rarely give orders
- Worry about whether they are liked by their crew members

Effective leadership takes many forms. The correct style for a particular situation or operation depends on the nature of the crew as well as the work it has to accomplish. For example, if the crew does not have enough experience for the job ahead, then a controller style may be appropriate. The controller style of leadership is also effective when jobs involve repetitive operations that require little decision-making.

However, if a worker's attitude is an issue, a directive style may be appropriate. In this case, providing the missing motivational factors may increase performance and result in the improvement of the worker's attitude. The directive style of leadership is also used when the work is of a creative nature, because brainstorming and exchanging ideas with such crew members can be beneficial.

The advisor style is effective with an experienced crew on a well-defined project. The company must give a crew leader sufficient authority to do the job. This authority must be commensurate with responsibility, and it must be made known to crew members when they are hired so that they understand who is in charge.

A crew leader must have an expert knowledge of the activities to be supervised in order to be effective. This is important because the crew members need to know that they have someone to turn to when they have a question or a problem, when they need some guidance, or when modifications or changes are warranted by the job.

Respect is probably the most useful element of authority. Respect usually derives from being fair to employees by listening to their complaints and suggestions, and by using incentives and rewards appropriately to motivate crew members. In addition, crew leaders who have a positive attitude and a favorable personality tend to gain the respect of their crew members as well as their peers. Along with respect comes a positive attitude from the crew members.

2.1.5 Ethics in Leadership

Crew leaders should maintain the highest standards of honesty and legality. Every day, the crew leader must make decisions that may have fairness and moral implications. When you make a dishonorable or unethical decision, it not only reflects on you, but also impacts other workers, peers, and the company as a whole.

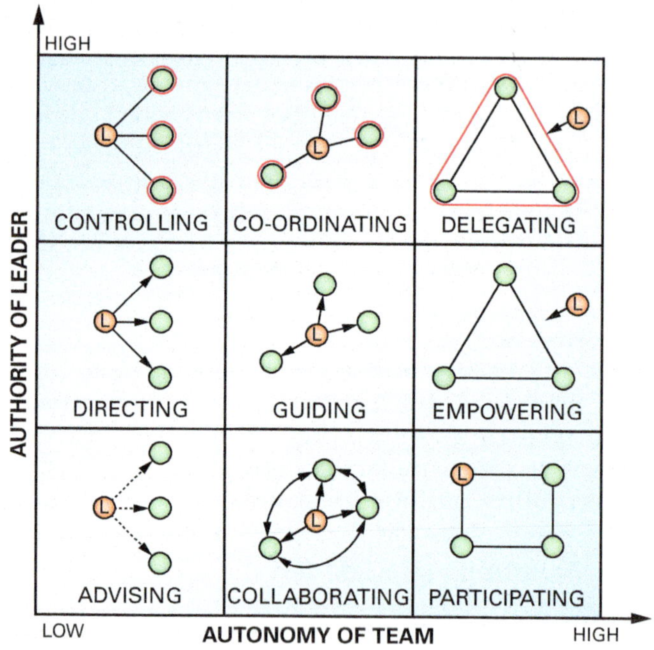

Figure 8 The authority-autonomy leadership style matrix.

There are three basic types of **ethics**:

- *Business or legal ethics* – Business, or legal, ethics concerns adhering to all laws and published regulations related to business relationships or activities.
- *Professional/balanced ethics* – Professional, or balanced, ethics relates to carrying out all activities in such a manner as to be honest and fair to everyone under one's authority.
- *Situational ethics* – Situational ethics pertains to specific activities or events that may initially appear to be a gray area. For example, you may ask yourself, "How will I feel about myself if my actions were going to be published in the newspaper or if I need to justify my actions to my family, friends, and colleagues? Would I still do the same thing?"

You will often find yourself in a situation where you will need to assess the ethical consequences of an impending decision. For example, if one of your crew members is showing symptoms of heat exhaustion, should you keep him working just because the superintendent says the project is behind schedule? If you are the only one aware that your crew did not properly erect the reinforcing steel, should you stop the pour and correct the situation? If supervision ever asks a crew leader to carry through on an unethical decision, it is up to that individual to inform the next-higher level of authority of the unethical nature of the issue. It is a sign of good character to refuse to carry out an unethical act.

"The right way is not always the popular and easy way. Standing for right when it is unpopular is a true test of character."
—Margaret Chase Smith

> **NOTE**
> Anyone who is aware of or knowingly a party to an illegal activity can face legal consequences simply due to inactivity, even if the activities are instigated by one's supervisor. However, a person of good character will never consider weighing the legal consequences against the ethics of one's actions.

2.2.0 Communication

Successful crew leaders learn to communicate effectively with people at all levels of the organization. In doing so, they develop an understanding of human behavior and acquire communication skills that enable them to understand and influence others.

There are many definitions of communication. Communication is the act of accurately and effectively conveying to or exchanging facts, feelings, and opinions with another person. Simply stated, communication is the process of exchanging information and ideas. Just as there are many definitions of communication, it also comes in many forms, including verbal, written, and nonverbal.

2.2.1 The Communication Process and Verbal Communication

There are two basic steps to clear communication, as illustrated in *Figure 9*. First, a sender sends a message (either verbal or written) to a receiver. When the receiver gets the message, he or she figures out what it means by listening or reading carefully. If anything is unclear, the receiver gives the sender feedback by asking the sender for more information.

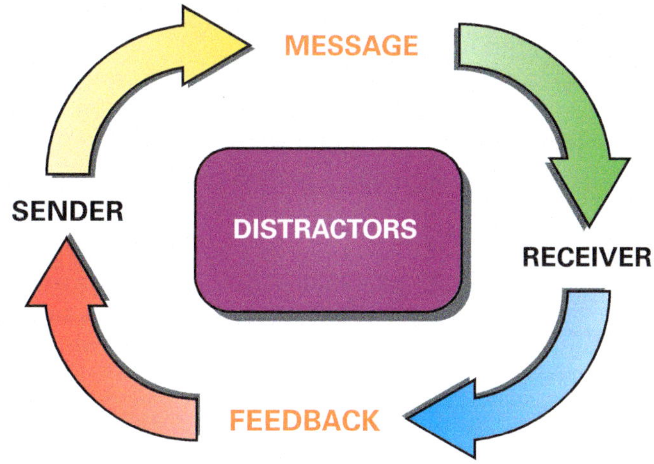

Figure 9 The communication process.

This process is called *two-way communication*, and it is the most effective way to make sure that everyone understands what's going on. This process sounds simple, so you may ask why good communication is so hard to achieve? When we try to communicate, a lot of things can get in the way. These communication obstacles and distractors are called *noise*.

The communication process illustrated in *Figure 9* consists of the following major components:

- *The Sender* – The sender is the person who creates and transmits the message. In verbal communication, the sender speaks the message aloud to the person(s) receiving the message. The sender must be sure to speak in a clear and concise manner easily understood by others. This isn't a natural skill; it takes practice. Some basic speaking tips are as follows:
  - Avoid talking with anything in your mouth (food, gum, etc.).
  - Avoid swearing, or other crude language, and acronyms.
  - Don't speak too quickly or too slowly. In extreme cases, people tend to focus on the rate of speech rather than the words themselves.

Did You Know?

Supervisor's Communication Breakdown

No, this isn't about failure of communication, but how a supervisor's daily communications break down into the various types, and how much time the supervisor spends doing each. Research shows that about 80 percent of the typical supervisor's day is spent communicating through writing, speaking, listening, or using body language. Of that time, studies suggest that approximately 20 percent of communication is in the written form, and 80 percent involves speaking or listening.

- Pronounce words carefully to prevent misunderstandings.
- Speak pleasantly and with enthusiasm. Avoid speaking in a harsh voice or in a monotone.

- *The Message* – The message is what the sender is attempting to communicate to the receiver(s). A message can be a set of directions, an opinion, or dealing with a personnel matter (praise, reprimand, etc.). Whatever its function, a message is an idea or fact that the sender wants the receiver to know.

  Before speaking, determine what must be communicated, and then take the time to organize what to say. Ensure that the message is logical and complete. Take the time to clarify your thoughts to avoid confusing the receiver. This also permits the sender to get to the point quickly.

  In delivering the message, the sender should assess the audience. It is important not to talk down to them. Remember that everyone, whether in a senior or junior position, deserves respect and courtesy. Therefore, the sender should use words and phrases that the receiver can understand and avoid technical language or slang if the receiver is unfamiliar with such terms. In addition, the sender should use short sentences, which gives the audience time to understand and digest one point or fact at a time.

- *The Receiver* – The receiver is the person who takes in the message. For the verbal communication process to be successful, it is important that the receiver understands the message as the speaker intended. Therefore, the receiver must avoid things that interfere with the delivery of the message. There are many barriers to effective listening, particularly on a busy construction job site. Some of these obstacles include the following:
  - Noise, visitors, cell phones, or other distractions
  - Preoccupation, being under pressure, or daydreaming
  - Reacting emotionally to what is being communicated
  - Thinking about how to respond instead of listening
  - Giving an answer before the message is complete
  - Personal bias/prejudice against the sender or the sender's communication style
  - Finishing the sender's sentence

  Some tips for overcoming these barriers include the following:
  - Take steps to minimize or remove distractions; learn to tune out your surroundings
  - Listen for key points
  - Take notes
  - Be aware of your personal biases, and try to stay open-minded; focus on the message, not the speaker
  - Allow yourself time to process your thoughts before responding
  - Let the sender communicate the message without interruption

  There are many ways for a receiver to show that he or she is actively listening to the message. The receiver can even accomplish this without saying a word. Examples include maintaining eye contact (*Figure 10*), nodding your head, and taking notes. Feedback can also provide an important type of response to a message.

- *Feedback* – Feedback is the receiver's communication back to the sender in response to the message. Feedback is a very important part of the communication process because it shows the sender how the receiver interpreted the message and that the receiver understood it as intended. In other words, feedback is a checkpoint to make sure the receiver and sender are on the same page.

  The receiver can paraphrase the message as a form of feedback to the sender. When paraphrasing, you use your own words to repeat the message. That way, you can show the sender that you interpreted the message correctly and could explain it to others if needed.

Figure 10 Eye contact is important when communicating face-to-face.

When providing feedback, the receiver can also request clarification or additional information, generally by asking questions.

One opportunity a crew leader can take to provide feedback is in the performance of crew evaluations. Many companies use formal evaluations on a yearly basis to assess workers' performance for pay increases and eligibility for advancement. These evaluations should not come as a once-a-year surprise to workers. An effective crew leader provides frequent performance feedback, and should ultimately summarize these communications in the annual performance evaluation. It is also important to stress the importance of self-evaluation with your crew.

2.2.2 Written or Visual Communications

Some communication must be written or visual. Written or visual communication includes messages or information documented on paper, or transmitted electronically using words, sketches, or other types of images.

Many messages on a job are in text form. Examples include weekly reports, requests for changes, purchase orders, and correspondence on a specific subject. These items are in writing because they can form a permanent record for business and historical purposes. Some communications on the job must be in the form of images. People are visual beings, and sketches, diagrams, graphs, and photos or videos can more effectively communicate some things. Examples include the plans or drawings used on a job, or a video of a work site accident scene.

> *"One picture is worth a thousand words."*
> *—San Antonio Express-News (1918)*

When writing or creating a visual message, it is best to assess the receiver (the reader) or the audience before beginning. The receiver must be able to read the message and understand the content; otherwise, the communication process will be unsuccessful. Therefore, the sender (the writer) should consider the actual meaning of words or diagrams and how others might interpret them. In addition, the writer should make sure that handwriting, if applicable, is legible.

Here are some basic tips for writing:

- Avoid emotion-packed words or phrases.
- Avoid making judgments unless asked to do so.
- Avoid using unfamiliar acronyms and **textspeak** (*Figure 11*).

Figure 11 Avoid textspeak for on-the-job written communications.

- Avoid using technical language or jargon unless the writer knows the receiver is familiar with the terms used.
- Be positive whenever possible.
- Be prepared to provide a verbal or visual explanation, if needed.
- Make sure that the document is legible.
- Present the information in a logical manner.
- Proofread your work; check for spelling, typographical, and grammatical errors—especially if relying on a spell-checker or an autocorrect typing feature.
- Provide an adequate level of detail.
- State the purpose of the message clearly.
- Stick to the facts.

Entire books are available that explain how to create effective diagrams and graphs. There are many factors that contribute to clear, concise, unambiguous images. The following are some basic tips for creating effective visual communications:

- Avoid making complex graphics; simplicity is better.
- Be prepared to provide a written or verbal explanation of the diagram, if needed.
- Ensure that the diagram is large enough to be useful.
- Graphs are better for showing numerical relationships than columns of numbers.
- Present the information in a logical order.
- Provide a **legend** if you include symbolic information.
- Provide an adequate level of detail and accuracy.

2.2.3 Nonverbal Communication

Unlike verbal or written communication, nonverbal communication doesn't involve spoken or written words, or images. Rather, nonverbal communication refers to things the speaker and the receiver see and hear in each other while communicating face to face. Examples include facial expressions, body movements, hand gestures, tone of voice, and eye contact.

Nonverbal communication can provide an external signal of an individual's inner emotions. It occurs at the same time as verbal communication. In most cases, the transmission of the sender and receiver nonverbal cues is totally unconscious.

People observe physical, nonverbal cues when communicating, making them just as important as words. Often, nonverbal signals influence people more than spoken words. Therefore, it's important to be conscious of nonverbal cues to avoid miscommunication based on your posture or expression. After all, these things may have nothing to do with the communication exchange; instead, they may be carrying over from something else going on in your day.

2.2.4 Communication Issues

It is important to note that everyone communicates differently; that is what makes us unique as individuals. As the diversity of the workforce changes, communication becomes even more challenging because the audience may include individuals from many different backgrounds. Therefore, it is necessary to assess the audience to determine how to communicate effectively with each individual.

The key to effective communication is to acknowledge that people are different, and to be able to adjust the communication style to meet the needs of the audience or the person on the receiving end of your message. This involves relaying the message in the simplest way possible, and avoiding the use of words that people may find confusing. Be aware of how you use technical language, slang, jargon, and words that have multiple meanings. Present the information in a clear, concise manner. Avoid rambling and always speak clearly, using good grammar.

> **NOTE**
>
> When working with bilingual workers, it is sometimes helpful to learn words and phrases commonly used in their language(s). They will respect you more for attempting to learn their language as they learn yours, and the effort will generate a closer comradery.

In addition, be prepared to communicate the message in multiple ways or adjust your level of detail or terminology to ensure that everyone understands the meaning as intended. For instance, a visual learner may need a map to comprehend directions. It may be necessary to overcome language barriers on the job site by using graphics or visual aids to relay the message. *Figure 12* shows how to tailor the message to the audience.

2.3.0 Motivation

The ability to motivate others is a key skill that leaders must develop. To motivate is to influence others to put forth more effort to accomplish something. For example, a crew member who skips breaks and lunch to complete a job on time is thought to be highly motivated, but crew members who do the bare minimum or just enough to keep their jobs are considered unmotivated.

A leader can **infer** an employee's motivation by observing various factors in the employee's performance. Examples of factors that may indicate a crew's motivation include the level or rate of unexcused absenteeism, the percentage of employee turnover, and the number of complaints, as well as the quality and quantity of work produced.

Different things motivate different people in different ways. Consequently, there is no single best approach to motivating crew members. It is important to recognize that what motivates one crew member may not motivate another. In addition, what works to motivate a crew member once may not motivate that same person again in the future.

Often, the needs and factors that motivate individuals are the same as those that create job satisfaction. These include the following:

- Accomplishment
- Change
- Job importance
- Opportunity for advancement
- Personal growth
- Recognition and praise
- Rewards

A crew leader's ability to satisfy these needs increases the likelihood of high morale within a crew. Morale refers to an individual's emotional outlook toward work and the level of satisfaction gained while performing the jobs assigned. High morale means that employees will be motivated to work hard, and they will have a positive attitude about coming to work and doing their jobs.

2.3.1 Accomplishment

Accomplishment refers to a worker's need to set challenging goals and achieve them. There is nothing quite like the feeling of achieving a goal, particularly a goal one never expected to accomplish in the first place.

Participant Exercise A, Part I

Read the following verbal conversations, and identify any problems:

Conversation I:

Judy: Hey, José...
José: What's up?
Judy: Has the site been prepared for the job trailer yet?
José: Job trailer?
Judy: The job trailer—it's coming in today. What time will the job site be prepared?
José: The trailer will be here about 1:00 PM.
Judy: The job site! What time will the job site be prepared?

Conversation II:

Jimar: Hey, Mike, I need your help.
Mike: What is it?
Jimar: You and Miguel go over and help Al's crew finish laying out the site.
Mike: Why me? I can't work with Miguel. He can't understand a word I say.
Jimar: Al's crew needs some help, and you and Miguel are the most qualified to do the job.
Mike: I told you, Jimar, I can't work with Miguel.

Conversation III:

Hiro: Hey, Jill.
Jill: Sir?
Hiro: Have you received the latest DOL, EEO requirement to be sure the OFCP administrator finds our records up to date when he reviews them in August?
Jill: DOL, EEO, and OFCP?
Hiro: Oh, and don't forget the MSHA, OSHA, and EPA reports are due this afternoon.
Jill: MSHA, OSHA, and EPA?

Conversation IV:

Lakeisha: Good morning, Roberto, would you do me a favor?
Roberto: Okay, Lakeisha. What is it?
Lakeisha: I was reading the concrete inspection report and found the concrete in Bays 4A, 3B, 6C, and 5D didn't meet the 3,000 psi strength requirements. Also, the concrete inspector on the job told me the two batches that came in today had to be refused because they didn't meet the slump requirements as noted on page 16 of the spec. I need to know if any placement problems happened on those bays, how long the ready mix trucks were waiting today, and what we plan to do to stop these problems in the future.

Participant Exercise A, Part II

Read the following written memos, and identify any problems:

Memo I:

Let's start with the transformer vault $285.00 due. For what you ask? Answer: practically nothing I admit, but here is the story. Paul the superintendent decided it was not the way good ole Comm Ed wanted it, we took out the ladder and part of the grading (as Paul instructed us to do) we brought it back here to change it. When Comm Ed the architect or DOE found out that everything would still work the way it was, Paul instructed us to reinstall the work. That is the whole story there is please add the $285.00 to my next payout.

____

____

____

____

____

____

____

____

Memo II:

Let's take rooms C 307-C-312 and C-313 we made the light track supports and took them to the job to erect them when we tried to put them in we found direct work in the way, my men spent all day trying to find out what to do so ask your Superintendent (Frank) he will verify seven hours pay for these men as he went back and forth while my men waited. Now the Architect has changed the system of hanging and has the gall to say that he has made my work easier, I can't see how. Anyway, we want an extra two (2) men for seven (7) hours for April 21 at $55.00 per hour or $385.00 on April 28th DOE Reference 197 finally resolved this problem. We will have no additional charges on DOE Reference 197, please note.

____

____

____

____

____

____

Crew leaders can help their crew members attain a sense of accomplishment by encouraging them to develop performance plans, such as goals for the year, the attaining of which the crew leader will consider later in performance evaluations. In addition, crew leaders can provide the support and tools (such as training and coaching) necessary to help their crew members achieve these goals.

2.3.2 Change

Change refers to an employee's need to have variety in work assignments. Change can keep things interesting or challenging, and prevent the boredom that results from doing the same task day after day with no variety. However, frequent or significant changes in work can actually have a negative impact on morale, since people often prefer some consistency and predictability in their lives.

2.3.3 Job Importance

Job importance refers to an employee's need to feel that his or her skills and abilities are valued and make a difference. Employees who don't feel valued tend to have performance and attendance issues. Crew leaders should attempt to make every crew member feel like an important part of the team, as if the job wouldn't be possible without their help.

2.3.4 Opportunity for Advancement

Opportunity for advancement refers to an employee's need to gain additional responsibility and develop new skills and abilities. It is important that employees know that they aren't limited to their current jobs. Let them know that they have a chance to grow with the company and to be promoted as recognition for excelling in their work when such opportunities occur.

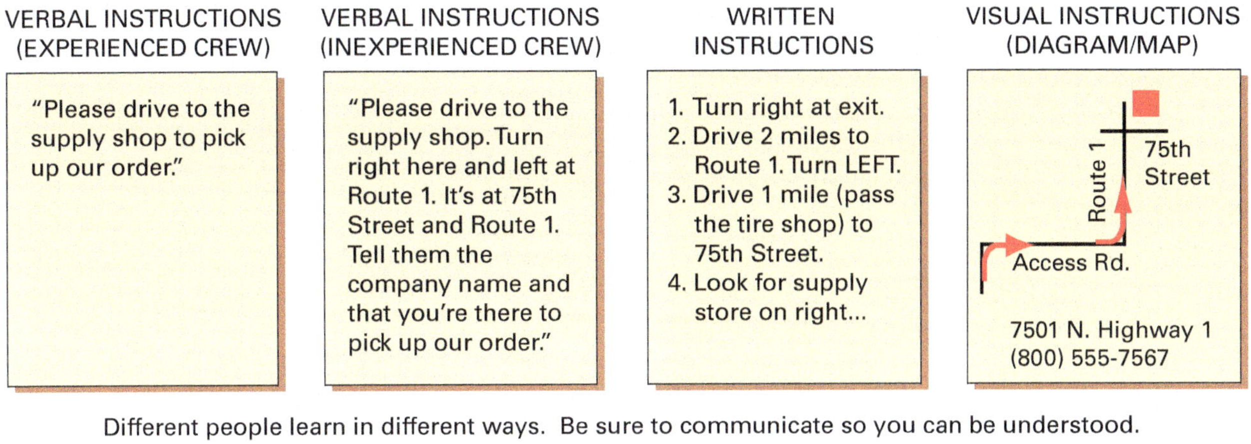

Figure 12 Tailor your message.

Effective leaders encourage each of their crew members to work to his or her full potential. In addition, they share information and skills with their employees to help them advance within the organization.

2.3.5 Recognition

Recognition and praise refer to the need to have good work appreciated, applauded, and acknowledged by others. You can accomplish this by simply thanking employees for helping on a project, or it can entail more formal praise, such as an award for Employee of the Month, or tangible or monetary awards. Some tips for giving recognition and praise include the following:

- Be available on the job site so that you have the opportunity to witness good work.
- Look for and recognize good work, and look for ways to praise it.
- Give recognition and praise only when truly deserved; people can quickly recognize insincere or artificial praise, and you will lose respect as a result.
- Acknowledge satisfactory performance, and encourage improvement by showing confidence in the ability of the crew members to do above-average work.

2.3.6 Personal Growth

Personal growth refers to an employee's need to learn new skills, enhance abilities, and grow as a person. It can be very rewarding to master a new competency on the job. Personal growth prevents the boredom associated with contemplation of working the same job indefinitely with no increase of technical knowledge, skills, or responsibility.

Crew leaders should encourage the personal growth of their employees as well as themselves. Learning should be a two-way street on the job site; crew leaders should teach their crew members and learn from them as well. In addition, crew members should be encouraged to learn from each other

Personal growth also takes place outside the workplace. Crew leaders should encourage their employees to acquire formal education applicable to their vocation as well as in subjects that will expand their understanding of the larger world. Encourage them to develop outside interests and ways to express their innate creativity.

2.3.7 Rewards

Rewards are additional compensation for hard work. Rewards can include an increase in a crew member's wages, or go beyond that to include bo-

Did You Know?

The Importance of Continuing Education

Education doesn't stop the day a person receives a diploma or certificate. It is a lifelong activity. Employers have long recognized and promoted continuing education as a factor in advancement, but it is essential to simply remaining in place as well. Regardless of what you do, new construction materials, methods, standards, and processes are constantly emerging. Those who don't make the effort to keep up will fall behind.

Also, studies have shown that a worker's lifetime income improved by 7–10 percent per year of community college. Consider continuing and vocational technical education (CTE and VTE) through local colleges.

nuses or other incentives. They can be monetary in nature (salary raises, holiday bonuses, etc.), or they can be nonmonetary, such as free merchandise (shirts, coffee mugs, jackets, merchant gift cards, etc.), or other prizes. Attendance at costly certification or training courses can be another form of reward.

2.3.8 Motivating Employees

To increase motivation in the workplace, crew leaders must individualize how they motivate different crew members. It is important that crew leaders get to know their crew members and determine what motivates them as individuals. Once again, as diversity increases in the workforce, this becomes even more challenging; therefore, effective communication skills are essential. Some tips for motivating employees include the following:

- Keep jobs challenging and interesting.
- Communicate your expectations. People need clear goals in order to feel a sense of accomplishment when they achieve them.
- Involve the employees. Feeling that their opinions are valued leads to pride in ownership and active participation.
- Provide sufficient training. Give employees the skills and abilities they need to be motivated to perform.
- Mentor the employees. Coaching and supporting employees boosts their self-esteem, their self-confidence, and ultimately their motivation.
- Lead by example. Employees will be far more motivated to follow someone who is willing to do the same things they are required to do.
- Treat employees well. Be considerate, kind, caring, and respectful; treat employees the way that you want to be treated.
- Avoid using scare tactics. Approaching your leadership responsibilities by threatening employees with negative consequences can result in higher employee turnover instead of motivation.
- Reward your crew for doing their best by giving them easier tasks from time to time. It is tempting to give your best employees the hardest or dirtiest jobs because you know they will do the jobs correctly.
- Reward employees for a job well done.
- Maintain a sense of humor, especially toward your own failings as a human being. No one is perfect. Your employees will appreciate that and be more motivated to be an encouragement to you.

2.4.0 Team Building

Organizations are making the shift from the traditional boss-worker mentality to one that promotes teamwork. The manager becomes the team leader, and the workers become team members. They all work together to achieve the common goals of the team.

There are several benefits associated with teamwork. These include the ability to complete complex projects more quickly and effectively, higher employee satisfaction, and reduced turnover.

2.4.1 Successful Teams

Successful teams consist of individuals who are willing to share their time and talents to reach a common goal—the goal of the team. Members of successful teams possess an "Us" or "We" attitude rather than an "I" and "You" attitude; they consider what's best for the team and put their egos aside.

Some characteristics of successful teams include the following:

- Everyone participates and every team member counts.
- All team members understand the goals of the team's work and are committed to achieve those goals.
- There is a sense of mutual trust and interdependence.
- The organization gives team members the confidence and means to succeed.
- They communicate.
- They are creative and willing to take risks.
- The team leader has strong people skills and is committed to the team.

2.4.2 Building Successful Teams

To be successful in the team leadership role, the crew leader should contribute to a positive attitude within the team. There are several ways in which the team leader can accomplish this. First, he or she can work with the team members to create a vision or purpose of what the team is to achieve. It is important that every team member is committed to the purpose of the team, and the team leader is instrumental in making this happen.

Within the construction industry, the company typically assigns a crew to a crew leader. However, it can be beneficial for the team leader to be involved in selecting the team members. The willingness of people to work on the team and the resources that they bring should be the key factors for selecting them for the team.

Participant Exercise B

You are the crew leader of a masonry crew. Sam Williams is the person whom the company holds responsible for ensuring that equipment is operable and distributed tothe jobs in a timely manner.

Occasionally, disagreements with Sam have resulted in tools and equipment arriving late. Sam, who has been with the company 15 years, resents having been placed in the job and feels that he outranks all the crew leaders.

Sam figured it was about time he talked with someone about the abuse certain tools and other items of equipment were receiving on some of the jobs. Saws were coming back with guards broken and blades chewed up, bits were being sheared in half, motor housings were bent or cracked, and a large number of tools were being returned covered with mud. Sam was out on your job when he observed a mason carrying a portable saw by the cord. As he watched, he saw the mason bump the swinging saw into a steel column. When the man arrived at his workstation, he dropped the saw into the mud.

You are the worker's crew leader. Sam approached as you were coming out of the work trailer. He described the incident. He insisted, as crew leader, you are responsible for both the work of the crew and how its members use company property. Sam concluded, "You'd better take care of this issue as soon as possible! The company is sick and tired of having your people mess up all the tools!"

You are aware that some members of your crew have been mistreating the company equipment.

1. How would you respond to Sam's accusations?

____

____

____

____

____

2. What action would you take regarding the misuse of the tools?

____

____

____

____

____

3. How can you motivate the crew to take better care of their tools? Explain.

____

____

____

____

____

When forming a new team, team leaders should do the following:

- Explain the purpose of the team. Team members need to know what they will be doing, how long they will be doing it (if they are temporary or permanent), and why they are needed.
- Help the team establish goals or targets. Teams need a purpose, and they need to know what it is they are responsible for accomplishing.
- Define team member roles and expectations. Team members need to know how they fit into the team and what they can expect to accomplish as members of the team.
- Plan to transfer a sense of responsibility to the team as appropriate. Teams should feel responsible for their assigned tasks. However, never forget that management will still hold the crew leader responsible for the crew's work.

"My model for business is The Beatles. They were four guys who kept each other's kind of negative tendencies in check. They balanced each other and the total was greater than the sum of the parts. That's how I see business: Great things in business are never done by one person, they're done by a team of people.

—Steve Jobs

2.4.3 Delegating

Once the various activities that make up the job have been determined, the crew leader must identify the person or persons who will be responsible for completing each activity. This requires that the crew leader be aware of the skills and abilities of the people on the crew. Then, the crew leader must put this knowledge to work in matching the crew's skills and abilities to accomplish the specific tasks needed to complete the job.

After matching crew members to specific activities, the crew leader must then delegate the assignments to the responsible person(s). Generally, when delegating responsibilities, the crew leader verbally communicates directly with the person who will perform or complete the activity.

When delegating work, remember to do the following:

- Delegate work to a crew member who can do the job properly. If it becomes evident that the worker doesn't perform to the desired standard, either teach the crew member to do the work correctly or turn it over to someone else who can (without making a public spectacle of the transfer).
- Make sure crew members understand what to do and the level of responsibility. Be clear about the desired results, specify the boundaries and deadlines for accomplishing the results, and note the available resources.
- Identify the standards and methods of measurement for progress and accomplishment, along with the consequences of not achieving the desired results. Discuss the task with the crew member and check for understanding by asking questions. Allow the crew member to contribute feedback or make suggestions about how to perform the task in a safe and quality manner.
- Give the crew member the time and freedom to get started without feeling the pressure of too much supervision. When making the work assignment, be sure to tell the crew member how much time there is to complete it, and confirm that this time is consistent with the job schedule.
- Examine and evaluate the result once a task is complete. Then, give the crew member some feedback as to how well the worker did the task. Get the crew member's comments. The information obtained from this is valuable and will enable the crew leader to know what kind of work to assign that crew member in the future. It will also provide a means of measuring the crew leader's own effectiveness in delegating work.

Be aware that there may be times when someone else in the company will issue written or verbal instructions to a crew member or to the crew without going through the crew leader. This kind of situation requires an extra measure of maturity and discretion on the part of the crew leader to first understand the circumstances, then establish an understanding with the responsible individual, if possible, to avoid work orders that circumvent the crew leader in the future.

2.4.4 Implementing Policies and Procedures

Every company establishes policies and procedures that crew leaders are expected to implement, and employees are expected to follow. Company policies and procedures are essentially guidelines for how the organization does business. They can also reflect organizational philosophies, such as putting safety first or making the customer the top priority. Examples of policies and procedures include safety guidelines, credit standards, and billing processes.

The following tips can help you effectively implement policies and procedures:

- Learn the purpose of each policy. This will help you follow the policy and apply it appropriately and fairly.
- If you're not sure how to apply a company policy or procedure, check the company manual or ask your supervisor.

> **NOTE**
> Try to obtain a supervisor's policy interpretation in writing or print out the email response so that you can append the decision to your copy of the company manual for future reference.

- Always follow company policies and procedures. Remember that they combine what's best for the customer and the company. In addition, they provide direction on how to handle specific situations and answer questions.

Crew leaders may need to issue orders to their crew members. An order is a form of communication that initiates, changes, or stops an activity. Orders may be general or specific, written or oral, and formal or informal. The decision of how an order will be issued is up to the crew leader, but the policies and procedures of the company may govern the choice.

When issuing orders, do the following:

- Make them as specific as possible. Avoid being general or vague unless it is impossible to foresee all the circumstances that could occur in carrying out the order.
- Recognize that it isn't necessary to write orders for simple tasks unless the company requires that supervisors write all orders.
- Write orders for more complex tasks, tasks that will take considerable time to complete, or that are permanent (standing) orders.
- Consider what is being said, the audience to whom it applies, and the situation under which it will be implemented to determine the appropriate level of formality for the order.

2.5.0 Making Decisions and Solving Problems

Decision making and problem solving and are a large part of every crew leader's daily work. They are a part of life for all supervisors, especially in fast-paced, deadlineoriented industries.

2.5.1 Decision Making Versus Problem Solving

Sometimes, the difference between decision making and problem solving isn't clear. Decision making refers to simply initiating an action, stopping one, or choosing an alternative course, as appropriate for the situation. Problem solving involves recognizing the difference between the way things are and the way things should be, then taking action to move toward the desired condition. The two activities are related because, to make a decision, you may have to use problem-solving techniques, just as solving problems requires making decisions.

2.5.2 Types of Decisions

Some decisions are routine or simple, and can be made based on past experiences. An example would be deciding how to get to work. If you've worked at the same place for a long time, you are already aware of the options for traveling to work (take the bus, drive a car, carpool with a co-worker, take a taxi, etc.). Based on past experiences with the options identified, you can make a decision about how best to get to work.

Other decisions are more difficult to make. These decisions require more careful thought about how to carry out an activity by using problem-solving techniques. An example is planning a trip to a new vacation spot. If you're not sure how to get there, where to stay, what to see, etc., one option is to research the area to determine the possible routes, hotel accommodations, and attractions. Then, you can make a decision about which route to take, what hotel to choose, and what sites to visit, without the benefit of direct experience. The Internet makes these tasks much easier, but the research still requires prioritizing what is most important to you.

2.5.3 Problem Solving

The ability to solve problems is an important skill in any workplace. It's especially important for craft professionals, whose workday is often not predictable or routine. This section provides a five-step process for solving problems, which you can apply to both workplace and personal issues.

Review the following steps and then see how you can be apply them to a job-related problem. Keep in mind that you can't solve a problem until everyone involved acknowledges the problem.

Step 1 *Define the problem.* This isn't as easy as it sounds. Thinking through the problem often uncovers additional problems. Also, drilling down to the facts of the problem may mean setting aside your own biases or presumptions toward the situation or the individuals involved.

Step 2 *Think about different ways to solve the problem.* There is often more than one solution to a problem, so you must think through each possible solution and pick the best one. The best solution might be taking parts of two different solutions and combining them to create a new solution.

Step 3 *Choose the solution that seems best, and make an action plan.* It is best to receive input both from those most affected by the problem, those who must correct the problem, and from those who will be most affected by any potential solution.

Step 4 *Test the solution to determine whether it actually works.* Many solutions sound great in theory but in practice don't turn out to be effective. On the other hand, you might discover from trying to apply a solution that it is acceptable with a little modification. If a solution doesn't work, think about how you could improve it, and then test your new plan.

Step 5 *Evaluate the process.* Review the steps you took to discover and implement the solution. Could you have done anything better? If the solution turns out to be satisfactory, you can add the solution to your knowledge base.

These five steps can be applied in specific situations. Read the following example situation, and apply the five-step problem-solving process to come up with a solution.

Example:

You are part of a team of workers assigned to a new shopping mall project. The project will take about 18 months to complete. The only available parking is half a mile from the job site. The crew must carry heavy toolboxes and safety equipment from their cars and trucks to the work area at the start of the day, and then carry them back at the end of their shifts. The five-step problem-solving process can be applied as follows:

Step 1 *Define the problem.* Workers are wasting time and energy hauling all their equipment to and from the work site.

Step 2 *Think about different ways to solve the problem.* Several workers have proposed solutions:

- Install lockers for tools and equipment closer to the work site.
- Have workers drive up to the work site to drop off their tools and equipment before parking.
- Bring in another construction trailer where workers can store their tools and equipment for the duration of the project.
- Provide a round-trip shuttle service to ferry workers and their tools.

> **NOTE**
> Each solution will have pros and cons, so it's important to receive input from the workers affected by the problem. For example, workers will probably object to any plan (like the drop-off plan) that leaves their tools vulnerable to theft.

Step 3 *Choose the solution that seems best, and make an action plan.* The work site superintendent doesn't want an additional trailer in your crew's area. The workers decide that the shuttle service makes the most sense. It should solve the time and energy problem, and workers can keep their tools with them. To put the plan into effect, the project supervisor arranges for a large van and driver to provide the shuttle service.

Step 4 *Test the solution to determine whether it actually works.* The solution works, but there is another problem. The workers' schedule has them all starting and leaving at the same time. There isn't enough room in the van for all the workers and their equipment. To solve this problem, the supervisor schedules trips spaced 15 minutes apart. The supervisor also adjusts worker schedules to correspond with the trips. That way, all the workers won't try to get on the shuttle at the same time.

Step 5 *Evaluate the process.* This process gave both management and workers a chance to express an opinion and discuss the various solutions. Everyone feels pleased with the process and the solution.

2.5.4 Special Leadership Problems

Because they are responsible for leading others, it is inevitable that crew leaders will encounter problems and be forced to make decisions about how to respond to the problem. Some problems will be relatively simple to resolve, like covering for a sick crew member who has taken a day off from work. Other problems will be complex and much more difficult to handle.

Some complex problems that are relatively common include the following:

- Inability to work with others
- Absenteeism and turnover
- Failure to comply with company policies and procedures

Inability to Work with Others – Crew leaders will sometimes encounter situations where an employee has a difficult time working with others on the crew. This could be a result of personality differences, gender or gender-identity prejudices, an inability to communicate, or some other cause. Whatever the reason, the crew leader must address the issue and get the crew working as a team.

The best way to determine the reason for why individuals don't get along or work well together is to talk to the parties involved. The crew leader should speak openly with the employee, as well as the other individual(s) to uncover the source of the problem and discuss its resolution.

After uncovering the reason for the conflict, the crew leader can determine how to respond. There may be a way to resolve the problem and get the workers communicating and working as a team again. On the other hand, there may be nothing the crew leader can do that will lead to a harmonious solution. In this case, the crew leader would need to either transfer one of the involved employees to another crew or have the problem crew member terminated. Resorting to this latter option should be the last measure and taken after discussing the matter with one's immediate supervisor or the company's Human Resources Department.

Absenteeism and Turnover – Absenteeism and turnover in the industry can delay jobs and cause companies to lose money. Absenteeism refers to workers missing their scheduled work time on a job. It has many causes, some of which cannot be helped. Sickness, family emergencies, and funerals are examples of unavoidable causes of worker absence. However, there are some causes of unexcused absenteeism that crew leaders can prevent.

The most effective way to control absenteeism is to make the company's policy clear to all employees. The policy should be explained to all new employees. This explanation should include the number of absences allowed and acceptable reasons for taking sick or personal days. In addition, all workers should know how to inform their crew leaders when they miss work and understand the consequences of exceeding the number of sick or personal days allowed.

Once crew leaders explain the policy on absenteeism to employees, they must be sure to implement it consistently and fairly. This makes employees more likely to follow it. However, if enforcement of the policy is inconsistent and the crew leader gives some employees exceptions, it won't be effective. Thus, the rate of absenteeism is likely to increase.

Despite having a policy on absenteeism, there will always be employees who are chronically late or miss work. In cases where an employee abuses the absenteeism policy, the crew leader should discuss the situation directly with the employee, ensure that they understand the policy, and insist that they comply with it. If the employee's behavior does not improve, disciplinary action will be in order.

Turnover refers to the rate at which workers leave a company and are replaced by others. Like absenteeism, there are some causes of turnover that cannot be prevented and others that can. For instance, an employee may find a job elsewhere earning twice as much money. However, crew leaders can prevent some employee turnover situations. They can work to ensure safe working conditions for their crew, treat their workers fairly and consistently, and help promote good working conditions. The key is communication and promoting the motivational factors discussed earlier. Crew leaders need to know the problems if they are going to be able to successfully resolve them.

Some major causes of employee turnover include the following:

- Unfair/inconsistent treatment by the immediate supervisor
- Unsafe project sites
- Lack of job security or opportunities for advancement

For the most part, the actions described for absenteeism are also effective for reducing turnover. Past studies have shown that maintaining harmonious relationships on the job site goes a long way in reducing both turnover and absenteeism. This requires effective and proactive leadership on the part of the crew leader.

Failure to Comply with Company Policies and Procedures – Policies are rules that define the relationship between the company, its employees, its clients, and its subcontractors. Procedures include the instructions for carrying out the policies. Some companies have policies that dictate dress codes. The dress code may be designed partly to ensure safety, and partly to define the image a company wants to project to the outside world.

Companies develop procedures to ensure that everyone who performs a task does it safely and efficiently. Many procedures directly relate to safety. A lockout/tagout (LOTO) procedure is an example. In this procedure, the company defines who may perform a LOTO, how to properly complete and remove a LOTO, and who has the authority to remove it. Workers who fail to follow the procedure endanger themselves, as well as their co-workers.

Companies typically have a policy on disciplinary action, which defines steps to take if an employee violates company policies or procedures. The steps range from counseling by a supervisor for the first offense, to a written warning and/or LOI, to dismissal for repeat offenses. This will vary from one company to another. For example, some companies will fire an employee for the first violation of a safety procedure with potential for loss of life or serious injury.

The crew leader has the first-line responsibility for enforcing company policies and procedures. The crew leader should take the time with a new crew member to discuss the policies and procedures and show the crew member how to access them. If a crew member shows a tendency to neglect a policy or procedure, it is up to the crew leader to counsel that individual. If the crew member continues to violate a policy or procedure, the crew leader has no choice but to refer that individual to the appropriate authority within the company for disciplinary action.

NOTE

When a crew shows a pattern of consistent policy violations, management will scrutinize the crew leader and potentially take disciplinary action if the crew leader does not handle the violations appropriately. The crew leader is responsible for all aspects of the crew, not just its work accomplishment on the job.

Participant Exercise C

Case I:

On the way over to the job trailer, you look up and see a piece of falling scrap heading for one of the laborers. Before you can say anything, the scrap material hits the ground about five feet in front of the worker. You notice the scrap is a piece of conduit. You quickly pick it up, assuring the worker you will take care of this matter.

Looking up, you see your crew on the third floor in the area from which the material fell. You decide to have a talk with them. Once on the deck, you ask the crew if any of them dropped the scrap. The men look over at Bob, one of the electricians in your crew. Bob replies, "I guess it was mine. It slipped out of my hand."

It is a known fact that the Occupational Safety and Health Administration (OSHA) regulations state that an enclosed chute of wood shall be used for material waste transportation from heights of 20 feet or more. It is also known that Bob and the laborer who was almost hit have been seen arguing lately.

1. Assuming Bob's action was deliberate, what action would you take?

   ____

   ____

   ____

2. Assuming the conduit accidentally slipped from Bob's hand, how can you motivate him to be more careful?

   ____

   ____

   ____

3. What follow-up actions, if any, should be taken relative to the laborer who was almost hit?

   ____

   ____

   ____

4. Should you discuss the apparent OSHA violation with the crew? Why or why not?

   ____

   ____

   ____

5. What acts of leadership would be effective in this case? To what leadership traits are they related?

   ____

   ____

   ____

Case II:

The company just appointed Antonio crew leader of a tile-setting crew. Before his promotion into management, he had been a tile setter for five years. His work had been consistently of superior quality.

Except for a little good-natured kidding, Antonio's co-workers had wished him well in his new job. During the first two weeks, most of them had been cooperative while Antonio was adjusting to his supervisory role.

At the end of the second week, a disturbing incident took place. Having just completed some of his duties, Antonio stopped by the job-site wash station. There he saw Steve and Ron, two of his old friends who were also in his crew, washing.

"Hey, Ron, Steve, you shouldn't be cleaning up this soon. It's at least another thirty minutes until quitting time," said Antonio. "Get back to your work station, and I'll forget I saw you here."

"Come off it, Antonio," said Steve. "You used to slip up here early on Fridays. Just because you have a little rank now, don't think you can get tough with us." To this Antonio replied, "Things are different now. Both of you get back to work, or I'll make trouble." Steve and Ron said nothing more, and they both returned to their work stations.

From that time on, Antonio began to have trouble as a crew leader. Steve and Ron gave him the silent treatment. Antonio's crew seemed to forget how to do the most basic activities. The amount of rework for the crew seemed to be increasing. By the end of the month, Antonio's crew was behind schedule.

1. How do you think Antonio should've handled the confrontation with Ron and Steve?

   ____

   ____

2. What do you suggest Antonio could do about the silent treatment he got from Steve and Ron?

   ____

   ____

3. If you were Antonio, what would you do to get your crew back on schedule?

   ____

   ____

4. What acts of leadership could be used to get the crew's willing cooperation?

   ____

   ____

5. To which leadership traits do they correspond?

   ____

   ____

Additional Resources

Construction Workforce Development Professional, NCCER. 2016. New York, NY: Pearson Education, Inc.

Mentoring for Craft Professionals, NCCER. 2016. New York, NY: Pearson Education, Inc.

It's Your Ship: Management Techniques from the Best Damn Ship in the Navy, Captain D. Michael Abrashoff, USN. 2012. New York City, NY: Grand Central Publishing.

Survival of the Fittest, Mark Breslin. 2005. McNally International Press.

The Definitive Book of Body Language: The Hidden Meaning Behind People's Gestures and Expressions, Barbara Pease and Allan Pease. 2006. New York City, NY: Random House / Bantam Books.

2.0.0 Section Review

1. A crew leader differs from a craftworker in that a crew leader _____.
   a. does not need direct experience in the job duties a craft professional typically performs
   b. can expect to oversee one or more workers in addition to performing typical craft duties
   c. is exclusively in charge of overseeing, since performing technical work isn't part of this role
   d. has no responsibility to be present on the job site

2. Feedback is important in verbal communication because it _____.
   a. requires the sender to repeat the message
   b. involves the receiver repeating back the message word for word
   c. informs the sender of how the message was received
   d. consists of a written analysis of the message

3. Once achieved, setting challenging goals for workers gives them a sense of _____.
   a. accomplishment
   b. entitlement
   c. persecution
   d. failure

4. Which of the following is *not* a characteristic of a successful team?
   a. Everyone participates and everyone counts.
   b. There is a sense of mutual trust.
   c. Members minimize communication with each other.
   d. The team leader is committed to the team.

5. Problem solving differs from decision making because it _____.
   a. involves finding an answer
   b. expresses an opinion
   c. involves starting or stopping an action
   d. separates facts from non-facts

SECTION THREE

3.0.0 SAFETY AND SAFETY LEADERSHIP

Objective

Identify a crew leader's typical safety responsibilities with respect to common safety issues, including awareness of safety regulations and the cost of accidents.

a. Explain how a strong safety program can enhance a company's success.
b. Explain the purpose of OSHA and describe the role of OSHA in administering worker safety.
c. Describe the role of employers in establishing and administering safety programs.
d. Explain how crew leaders are involved in administering safety policies and procedures.

Trade Terms

Intangible: Not touchable, material, or measureable; lacking a physical presence.

Lethargy: Sluggishness, slow motion, lack of activity, or a lack of enthusiasm.

Negligence: Lack of appropriate care when doing something, or the failure to do something, usually resulting in injury to an individual or damage to equipment.

Safety data sheets (SDS): Documents listing information about a material or substance that includes common and proper names, chemical composition, physical forms and properties, hazards, flammability, handling, and emergency response in accordance with national and international hazard communication standards. Also called material safety data sheets (MSDS).

Businesses lose millions of dollars every year because of on-the-job accidents. Work-related injuries, sickness, and fatalities have caused untold suffering for workers and their families. Resulting project delays and budget overruns can cause huge losses for employers, and work-site accidents damage the overall morale of the crew.

Craft professionals routinely face hazards. Examples of these hazards include falls from heights, working on scaffolds, using cranes in the presence of power lines, operating heavy machinery, and working on electrically-powered or pressurized equipment. Despite these hazards, experts believe that applying preventive safety measures could drastically reduce the number of accidents.

Did You Know?

The Fatal Four

When OSHA inspects a job site, they focus on the types of safety hazards that are most likely to cause serious and fatal injuries. These hazards result in the following most common categories of injuries:

- Falls from elevations
- Struck-by hazards
- Caught-in/between hazards
- Electrical-shock hazards

As a crew leader, one of your most important tasks is to enforce the company's safety program and make sure that all workers are performing their tasks safely. To be successful, the crew leader should do the following:

- Be aware of the human and monetary costs of accidents.
- Understand all federal, state, and local governmental safety regulations applicable to your work.
- Be the most visible example of the best safe work practices.
- Be involved in training workers in safe work methods.
- Conduct training sessions.
- Get involved in safety inspections, accident investigations, and fire protection and prevention.

"Example is not the main thing in influencing others. It is the only thing."
—Albert Schweitzer

Crew leaders are in the best position to ensure that their crew members perform all jobs safely. Providing employees with a safe working environment by preventing accidents and enforcing safety standards will go a long way towards maintaining the job schedule and enabling a job's completion on time and within budget.

3.1.0 The Impact of Accidents

Each day, workers in construction and industrial occupations face the risk of falls, machinery accidents, electrocutions, and other potentially fatal hazards. The National Institute of Occupational Safety and Health (NIOSH) statistics show that roughly 1,000 construction workers are killed on the job each year, which is more than any other industry. Falls are the leading cause of deaths in the construction industry through accident or negligence, accounting for over 60 percent of the fatalities in recent years. Nearly half of the fatal falls occurred from roofs, scaffolds, or ladders. Roofers, structural metal workers, and painters are at the greatest risk of fall fatalities (*Figure 13*).

In addition to the number of fatalities that occur each year, there are a staggering number of work-related injuries. In 2015, for example, almost 200,000 job-related injuries occurred in the construction industry. NIOSH estimates that the total cost of fatal and non-fatal injuries in the construction industry represents about 15 percent of the costs for all private industry. The main causes of injuries on construction sites include falls, electrocution, fires, and mishandling of machinery or equipment. According to NIOSH, back injuries are the leading health-related problem in workplaces.

3.1.1 Cost of Accidents

Occupational accidents cost roughly $250 billion or more every year. These costs affect the individual employee, the company, and the construction industry as a whole.

Organizations encounter both direct and indirect costs associated with workplace accidents. Direct costs are the money companies must pay out to workers' compensation claims and sick pay; indirect costs are all the other tangible and intangible things and costs a company must account for as the result of a worker's injury or death. To compete and survive, companies must control these as well as all other employment-related

Figure 13 Falls are the leading cause of deaths and injuries in construction.

costs. There are many costs involved with workplace accidents. A company can insure some of these costs, but not others.

Insured costs – Insured costs are those costs either paid directly or reimbursed by insurance carriers. Insured costs related to injuries or deaths include the following:

- Compensation for lost earnings (known as *worker's comp*)
- Funeral charges
- Medical and hospital costs
- Monetary awards for permanent disabilities
- Pensions for dependents
- Rehabilitation costs

Insurance premiums or charges related to property damages include the following:

- Fire or other safety-related peril
- Structural loss; material and equipment loss or damage
- Loss of business use and occupancy
- Public liability
- Replacement cost of equipment, material, and structures

Uninsured costs – The relative direct and indirect costs of accidents are comparable to the visible and hidden portions of an iceberg, as shown in *Figure 14*. The tip of the iceberg represents direct costs, which are the visible costs. Not all of these are covered by insurance. The more numerous indirect costs aren't readily measurable, but they can represent a greater financial burden than the direct costs.

Uninsured costs from injuries or deaths include the following:

- First aid expenses
- Transportation costs
- Costs of investigations
- Costs of processing reports
- Down time on the job site
- Costs to train replacement workers

Did You Know?

The Costs of Negligence

If you receive an injury as the result of a workplace accident, and a completed investigation shows that your injuries were due to your own negligence, you can't sue your employer for damages. In addition, workers' compensation insurance companies may decline to pay claims for negligent employees who are injured or killed.

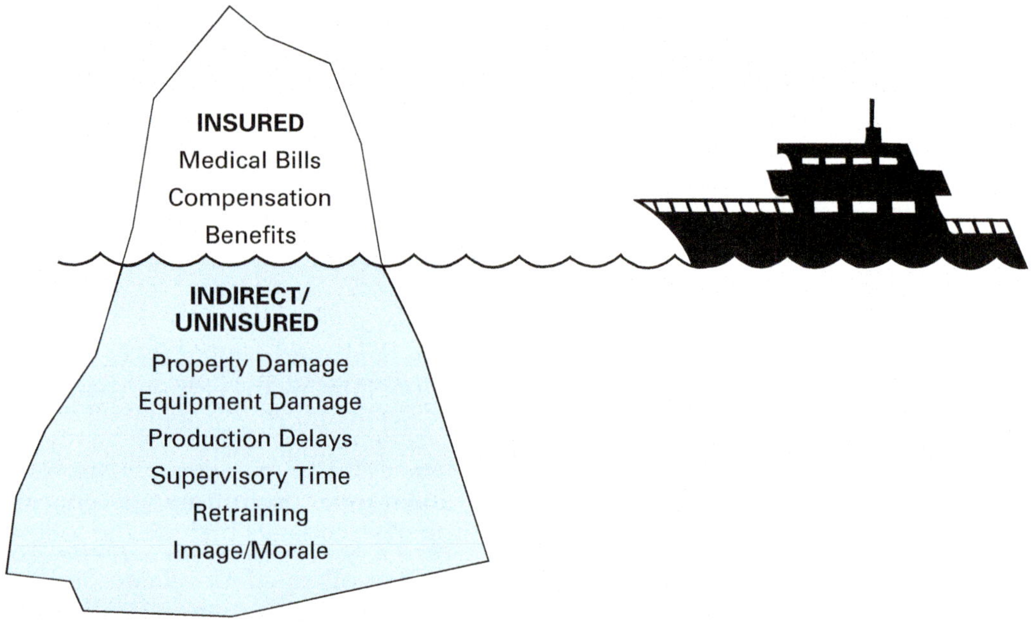

Figure 14 Costs associated with accidents.

Uninsured costs related to wage losses include the following:

- Idle time of workers whose work is interrupted
- Time spent cleaning the accident area
- Time spent repairing damaged equipment
- Time lost by workers receiving first aid
- Costs of training injured workers in a new career

Uninsured costs related to production losses include the following:

- Product spoiled by accident
- Loss of skill and experience; worker replacement
- Lowered production capacity
- Idle machine time due to lack of qualified operators

Associated costs may include the following:

- Difference between actual losses and amount recovered
- Costs of rental equipment used to replace damaged equipment
- Costs of inexperienced temp or permanent new workers used to replace injured workers
- Wages or other benefits paid to disabled workers
- Overhead costs while production is stopped
- Impact on schedule
- Loss of client bonus or payment of forfeiture for delays

Uninsured costs related to off-the-job activities include the following:

- Time spent on ensuring injured workers' welfare
- Loss of skill and experience of injured workers
- Costs of training replacement workers

Uninsured costs related to intangible factors include the following:

- Increased labor conflict
- Loss of bid opportunities because of poor safety records
- Loss of client goodwill
- Lowered employee morale
- Unfavorable public relations

3.2.0 OSHA

To reduce safety and health risks and the number of injuries and fatalities on the job, the federal government has enacted laws and regulations, including the Occupational Safety and Health Act of 1970 (OSH Act of 1970). This law created the Occupational Safety and Health Administration (OSHA), which is part of the US Department of Labor. OSHA also provides education and training for employers and workers. Through the administration of OSHA, the US Congress seeks "to assure so far as possible every working man and woman in the Nation safe and healthful working conditions and to preserve our human resources…" (OSH Act of 1970, Section 2[b]).

To promote a safe and healthy work environment, OSHA issues standards and rules for working conditions, facilities, equipment, tools, and work processes. It does extensive research into occupational accidents, illnesses, injuries, and deaths to reduce the number of occurrences and adverse effects. In addition, OSHA regulatory agencies conduct workplace inspections to ensure that companies follow the standards and rules.

To enforce OSHA regulations, the government has granted regulatory agencies the right to enter public and private properties to conduct workplace safety investigations. The agencies also have the right to take legal action if companies are not in compliance with the Act. These regulatory agencies employ OSHA Compliance Safety and Health Officers (CSHO), who are experts in the occupational safety and health field. The CSHOs are thoroughly familiar with OSHA standards and recognize safety and health hazards.

States with their own occupational safety and health programs conduct inspections by enlisting the services of qualified state CSHOs.

Companies are inspected for a multitude of reasons. They may be randomly selected, or they may be chosen as a result of employee complaints, a report of an imminent danger, or major accidents/fatalities that have occurred.

OSHA has established significant monetary fines for the violation of its regulations. *Table 1* lists the penalties as of 2016. In some cases, OSHA will hold a superintendents or crew leaders personally liable for repeat violations as well. In addition to the fines, there are possible criminal charges for willful violations resulting in death or serious injury. The attitude of the employer and their safety history can have a significant effect on the outcome of a case.

3.3.0 Employer Safety Responsibilities

Each employer must set up a safety and health program to manage workplace safety and health and to reduce work-related injuries, illnesses, and fatalities. The program must be appropriate for the conditions of the workplace. It should consider the number of workers employed and the hazards they face while at work.

To be successful, the safety and health program must have management, leadership, and employee participation. In addition, training and informational meetings play an important part in effective programs. Being consistent with safety policies is the key. Regardless of the employer's responsibility, however, the individual worker is ultimately responsible for his or her own safety.

Table 1 OSHA Penalties for violations established in 2016

Type of Violation	Maximum Penalty
Serious Other-Than-Serious Posting Requirements	$12,471 per violation
Failure to Abate	$12,471 per day
Willful or Repeated Violation	$124,709 per violation

Did You Know?

State OSHA Programs

Nearly half the states in the United States and some of its territories have state-run OSHA programs. These programs are set up under federal OSHA guidelines and must establish job health and safety standards that are at least as effective as the federal standards. The states/territories have the option of adopting more stringent standards or setting standards for hazards not addressed in the federal program. Of the 22 states and territories with non-federal OSHA programs, six of them limit their coverage to state and local government employees.

3.3.1 Safety Program

The crew leader plays a key role in the successful implementation of the safety program. The crew leader's attitude toward the program sets the standard for how crew members view safety. Therefore, the crew leader should follow all program guidelines and require crew members to do the same.

Safety programs should consist of the following:

- Safety policies and procedures
- Safety information and training
- Posting of safety notices
- Hazard identification, reporting, and assessment
- Safety record system
- Accident reporting and investigation procedures
- Appropriate discipline for not following safety procedures

3.3.2 Safety Policies and Procedures

Employers are responsible for following OSHA and state safety standards. Usually, they incorporate federal and state OSHA regulations into a safety policies and procedures manual. Employees receive such a manual when they are hired.

During orientation, appropriate company staff should guide the new employees through the general sections of the safety manual and the sections that have the greatest relevance to their job.

If the employee can't read, the employer should have someone read it to the employee and answer any questions that arise. The employee should then sign a form stating understanding of the information.

It isn't enough to tell employees about safety policies and procedures on the day they are hired and then never mention them again. Rather, crew leaders should constantly emphasize and reinforce the importance of adhering to all safety policies and procedures. In addition, employees should play an active role in determining job safety hazards and find ways to prevent and control hazards.

3.3.3 Hazard Identification and Assessment

Safety policies and procedures should be specific to the company. They should clearly present the hazards of the job and provide the means to report hazards to the proper level of management without prejudice to the individual doing the reporting. Crew leaders should also identify and assess hazards to which employees are exposed. They must also assess compliance with federal and state OSHA standards.

To identify and assess hazards, OSHA recommends that employers conduct periodic and random inspections of the workplace, monitor safety and health information logs, and evaluate new equipment, materials, and processes for potential hazards before they are used.

"You get what you inspect, not what you expect."
—Anonymous

Crew leaders and workers play important roles in identifying and reporting hazards. It is the crew leader's responsibility to determine what working conditions are unsafe and to inform employees of hazards and their locations. In addition, they should encourage their crew members to tell them about hazardous conditions. To accomplish this, crew leaders must be present and available on the job site.

The crew leader also needs to help the employee be aware of and avoid the built-in hazards to which craft professionals are exposed. Examples include working at elevations, working in confined spaces such as tunnels and underground vaults, on caissons, in excavations with earthen walls, and other naturally-dangerous projects. In addition, the crew leader can take safety measures, such as installing protective railings to prevent workers from falling from buildings, as well as scaffolds, platforms, and shoring.

3.3.4 Safety Information and Training

The employer must provide periodic information and training to new and long-term employees. This happens as often as necessary so that all employees receive adequate training. When safety and health information changes or workplace conditions create new hazards, the company must then provide special training and informational sessions. It is important to note that the company must present safety-related information in a manner that each employee will understand.

When a crew leader assigns an inexperienced employee a new task, the crew leader must ensure that the employee can do the work in a safe manner. The crew leader can accomplish this by providing safety information or training for groups or individuals.

When assigning an inexperienced employee a new task, do the following:

- Define the task.
- Explain how to do the task safely.
- Explain what tools and equipment to use and how to use them safely.
- Identify the necessary personal protective equipment and train the employee in its use.
- Explain the nature of the hazards in the work and how to recognize them.
- Stress the importance of personal safety and the safety of others.
- Hold regular safety training sessions with the crew's input.
- Review **safety data sheets (SDS)** that may be applicable.

3.3.5 Safety Record System

OSHA regulations (29 *CFR* 1904) require that employers keep records of hazards identified and document the severity of the hazard. The information should include the likelihood of employee exposure to the hazard, the seriousness of the harm associated with the hazard, and the number of exposed employees.

In addition, the employer must document the actions taken or plans for action to control the hazards. While it is best to take corrective action immediately, it is sometimes necessary to develop a plan to set priorities and deadlines and track progress in controlling hazards.

Employers who are subject to the recordkeeping requirements of the Occupational Safety and Health Act of 1970 must maintain records of all recordable occupational injuries and illnesses. The following are some OSHA forms that should be used for this recordkeeping:

- OSHA Form 300, *Log of Work-Related Injuries and Illnesses*
- OSHA Form 300A, *Summary of Work-Related Injuries and Illnesses*
- OSHA Form 301, *Injury and Illness Incident Report*

These three OSHA forms are included in the *Appendix* at the end of this module. Note that crew leaders directly handle the OSHA Form 301.

An SDS provides both workers and emergency personnel with the proper procedures for handling or working with a substance that may be dangerous. The document will include information such as physical data (melting point, boiling point, flash point, etc.), toxicity, health effects, first aid, reactivity, storage, disposal, protective equipment required for handling, and spill/leak procedures. These sheets are of particular use if a spill, fire, or other accident occurs.

Companies not exempted by OSHA must maintain required safety logs and retain them for 5 years following the end of the calendar year to which they relate. Logs must be available (normally at the company offices) for inspection and copying by representatives of the Department of Labor, the Department of Health and Human Services, or states given jurisdiction under the Act. Employees, former employees, and their representatives may also review these logs.

3.3.6 Accident Investigation

Employees must know from their training to immediately report any unusual event, accident, or injury. Policies and definitions of what these incidents consist of should be included in safety manuals. In the event of an accident, the employer is required to investigate the cause of the accident and determine how to avoid it in the future.

According to OSHA regulations, the employer must investigate each work-related death, serious injury or illness, or incident having the potential to cause death or serious physical harm. The employer should document any findings from the investigation, as well as the action plan to

Summaries of Work-Related Injuries and Illnesses

Most companies with 11 or more employees must post an OSHA Form 300A, *Summary of Work-Related Injuries and Illnesses,* between February 1 and April 30 of each year. Employees have the right to review this form. Check your company's policies regarding this and the related OSHA forms.

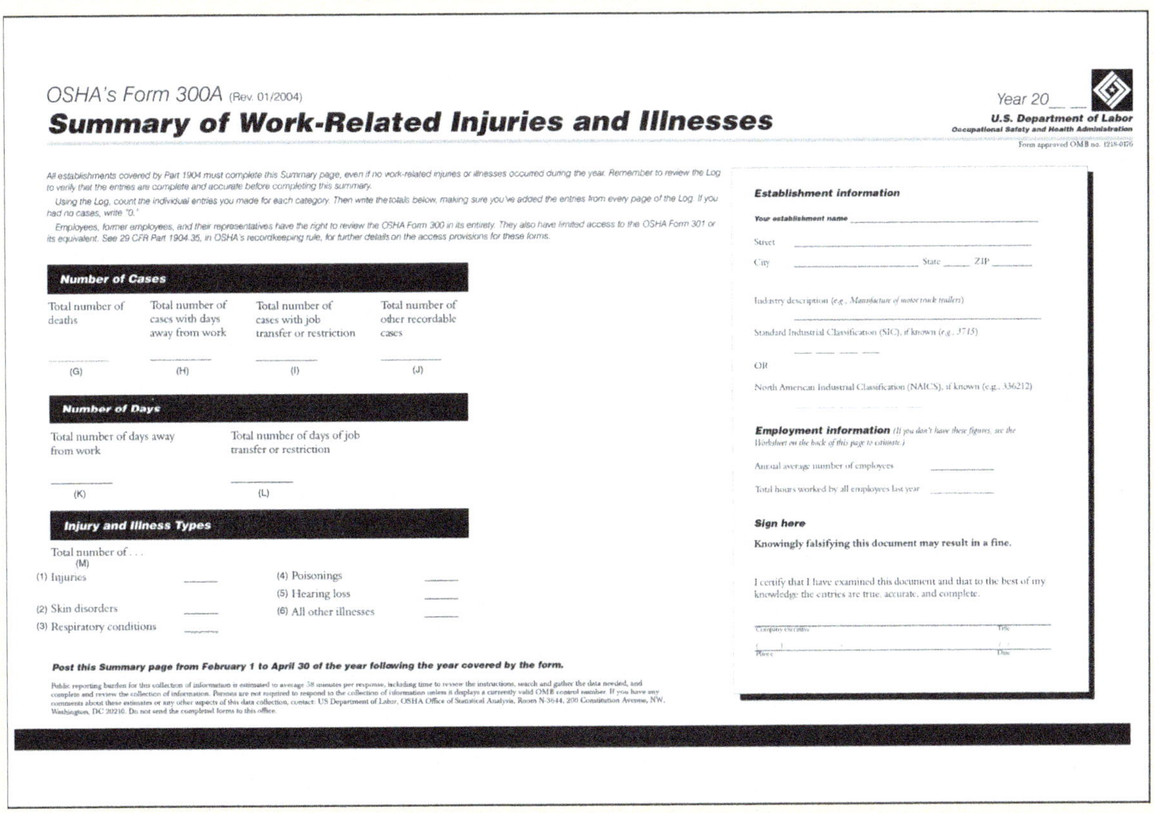

OSHA's Form 300A (Rev. 01/2004)

Summary of Work-Related Injuries and Illnesses

Year 20__ __

U.S. Department of Labor
Occupational Safety and Health Administration
Form approved OMB no. 1218-0176

All establishments covered by Part 1904 must complete this Summary page, even if no work-related injuries or illnesses occurred during the year. Remember to review the Log to verify that the entries are complete and accurate before completing this summary.

Using the Log, count the individual entries you made for each category. Then write the totals below, making sure you've added the entries from every page of the Log. If you had no cases, write "0."

Employees, former employees, and their representatives have the right to review the OSHA Form 300 in its entirety. They also have limited access to the OSHA Form 301 or its equivalent. See 29 CFR Part 1904.35, in OSHA's recordkeeping rule, for further details on the access provisions for these forms.

Number of Cases

Total number of deaths ______ (G)
Total number of cases with days away from work ______ (H)
Total number of cases with job transfer or restriction ______ (I)
Total number of other recordable cases ______ (J)

Number of Days

Total number of days away from work ______ (K)
Total number of days of job transfer or restriction ______ (L)

Injury and Illness Types

Total number of . . .
(M)
(1) Injuries ______
(2) Skin disorders ______
(3) Respiratory conditions ______
(4) Poisonings ______
(5) Hearing loss ______
(6) All other illnesses ______

Post this Summary page from February 1 to April 30 of the year following the year covered by the form.

Public reporting burden for this collection of information is estimated to average 58 minutes per response, including time to review the instructions, search and gather the data needed, and complete and review the collection of information. Persons are not required to respond to the collection of information unless it displays a currently valid OMB control number. If you have any comments about these estimates or any other aspects of this data collection, contact: US Department of Labor, OSHA Office of Statistical Analysis, Room N-3644, 200 Constitution Avenue, NW, Washington, DC 20210. Do not send the completed forms to this office.

Establishment information

Your establishment name ______
Street ______
City ______ State ______ ZIP ______

Industry description (*e.g., Manufacture of motor truck trailers*) ______
Standard Industrial Classification (SIC), if known (*e.g., 3715*) ______
OR
North American Industrial Classification (NAICS), if known (e.g., 336212) ______

Employment information *(If you don't have these figures, see the Worksheet on the back of this page to estimate.)*

Annual average number of employees ______
Total hours worked by all employees last year ______

Sign here

Knowingly falsifying this document may result in a fine.

I certify that I have examined this document and that to the best of my knowledge the entries are true, accurate, and complete.

Company executive ______ Title ______
Phone ( ) ______ Date / /

prevent future occurrences. The company should complete these actions immediately, with photos or video if possible. It's important that the investigation uncover the root cause of the accident to avoid similar incidents in the future. In many cases, the root cause was a flaw in the system that failed to recognize the unsafe condition or the potential for an unsafe act (*Figure 15*).

3.4.0 Leader Involvement in Safety

To be an effective, you must be actively involved in your company's safety program. Crew leader involvement includes conducting frequent safety training sessions and inspections, promoting first aid, and fire protection and prevention, preventing substance abuse on the job, and investigating accidents. Most importantly, crew leaders must practice safety at all times.

3.4.1 Safety Training Sessions

A safety training session may be a brief, informal gathering of a few employees or a formal meeting with instructional videos and talks by guest speakers. The size of the audience and the topics addressed determine the format of the meeting. You should plan to conduct small, informal safety sessions weekly.

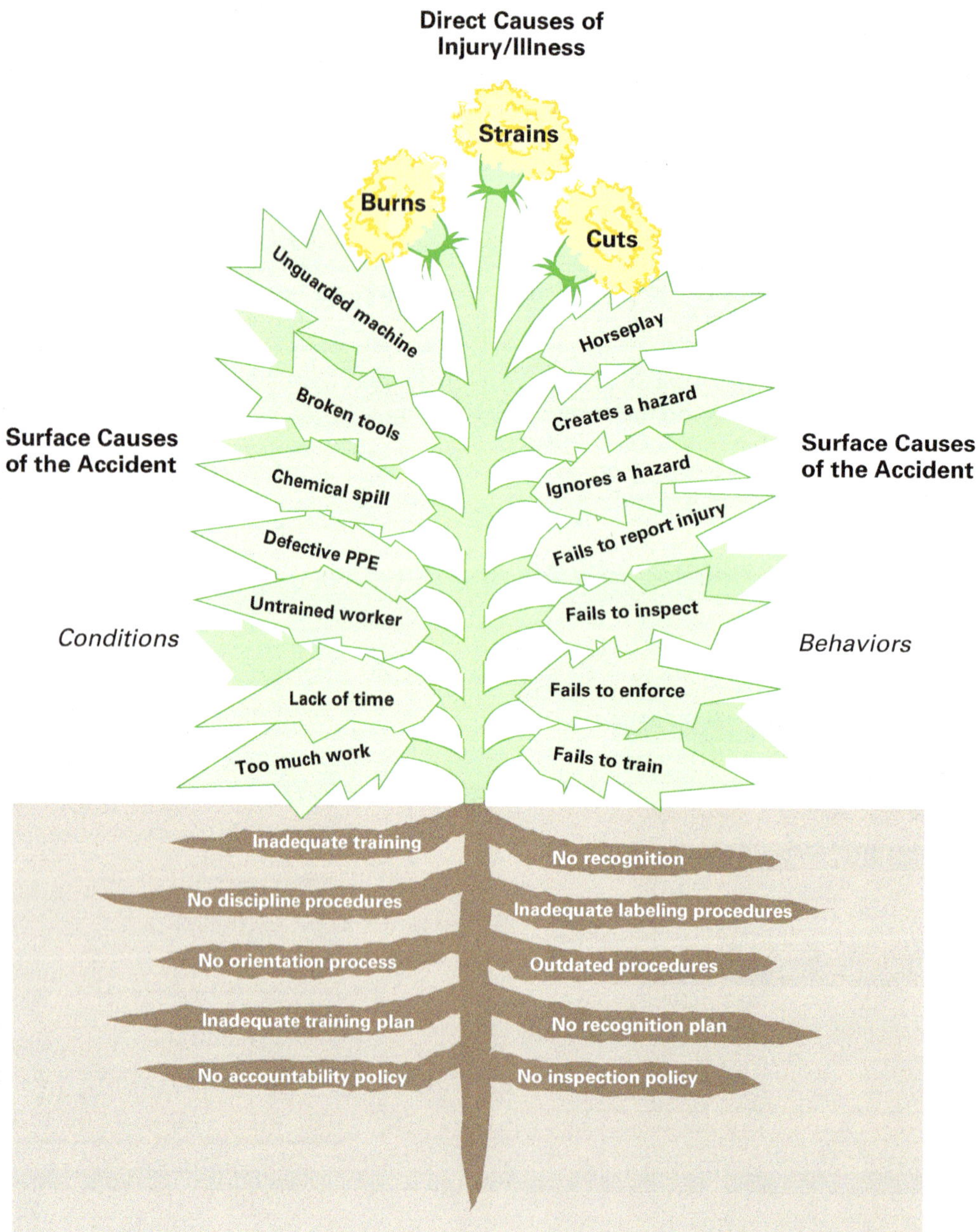

Figure 15 Root causes of accidents.

OSHA Accident Notification Requirements

There are urgent reporting requirements by the employer to OSHA for the following cases:

1. Within 8 hours: A work-related fatality.
2. Within 24 hours:
   - Work-related accident that resulted in an in-patient admission of one or more employees
   - Work-related amputation
   - Work-related loss of an eye

You should also plan safety training sessions in advance, and you should communicate the information to all affected employees. In addition, the topics covered in these training sessions should be timely and practical. Keep a log of each safety session signed by all attendees. It must be maintained as a record and available for inspection. It is advisable to attach a copy of a summary of the training session to the attendance record so you can keep track of what topics you covered and when.

3.4.2 Inspections

Crew leaders must make routine, frequent inspections to prevent accidents from happening. They must also take steps to avoid accidents. For that purpose, they need to inspect the job sites where their workers perform tasks. It's advisable to do these inspections before the start of work each day and during the day at random times.

Crew leaders must protect workers from existing or potential hazards in their work areas. Crews are sometimes required to work in areas controlled by other contractors. In these situations, the crew leader must maintain control over the safety exposure of the crew. If hazards exist, the crew leader should immediately bring the hazards to the attention of the contractor at fault, their superior, and the person responsible for the job site.

Crew leader inspections are only valuable if follow-up action corrects potential hazards. Therefore, crew leaders must be alert for unsafe and negligent acts on their work sites. When an employee performs an unsafe action, the crew leader must explain to the employee why the act was unsafe, tell that employee to not do it again, and request cooperation in promoting a safe working environment. The crew leader must document what happened and what the employee was told to do to correct the situation. It is then important that crew leaders follow up to make certain the employee is complying with the safety procedures. Never allow a safety violation to go uncorrected. There are three courses of action that you, as a crew leader, can take in an unsafe work site situation:

- Get the appropriate party to correct the problem.
- Fix the problem yourself.
- Refuse to have the crew work in the area until the responsible party corrects the problem.

"As soon as you see a mistake and don't fix it, it becomes your mistake."
—Anonymous

3.4.3 First Aid

The primary purpose of first aid is to provide immediate and temporary medical care to employees involved in accidents, as well as employees experiencing non-work-related health emergencies, such as chest pain or breathing difficulty. To meet this objective, every crew leader should be aware of the location and contents of first aid kits available on the job site. Emergency numbers should be posted in the job trailer. In addition, OSHA requires that at least one person trained in first aid be present at the job site at all times. It's also advisable, but not required, that someone on site should have cardiopulmonary resuscitation (CPR) training.

NOTE

CPR certifications must be renewed every two years.

The victim of an accident or sudden illness at a job site may be harder to aid than elsewhere since the worker may be at a remote location. The site may be far from a rescue squad, fire department, or hospital, presenting a problem in the rescue and transportation of the victim to a hospital. The worker may also have received an injury from falling rock or other materials, so immediate special rescue equipment or first-aid techniques are often needed.

Employer benefits of having personnel trained in first aid at job sites include the following:

- The immediate and proper treatment of minor injuries may prevent them from developing into more serious conditions. Thus, these precautions can eliminate or reduce medical expenses, lost work time, and sick pay.

- It may be possible to determine if the injured person requires professional medical attention.
- Trained individuals can save valuable time preparing the inured for treatment for when professional medical care arrives. This service increases the likelihood of saving a life.

The American Red Cross, Medic First Aid, and the United States Bureau of Mines provide basic and advanced first aid courses at nominal costs. These courses include both first aid and CPR. The local area offices of these organizations can provide further details regarding the training available.

3.4.4 Fire Protection and Prevention

Fires and explosions kill and injure many workers each year, so it is important that crew leaders understand and practice fire-prevention techniques as required by company policy.

The need for protection and prevention is increasing as manufacturers introduce new building materials. Some building materials are highly flammable. They produce great amounts of smoke and gases, which cause difficulties for fire fighters, and can quickly overcome anyone present. Other materials melt when they burn and may puddle over floors, preventing fire-fighting personnel from entering areas where this occurs.

OSHA has specific standards for fire safety. Employers are required to provide proper exits, fire-fighting equipment, and employee training on fire prevention and safety. For more information, consult OSHA guidelines (available at **www.osha.gov**).

3.4.5 Substance Abuse

Substance abuse is a continuing problem in the workplace. Substance abuse is the inappropriate overuse of drugs and chemicals, whether they are legal or illegal. All substance abuse results in some form of mental, sensory, or physical impairment. Some people use illegal "street drugs", such as cocaine or crystal meth. Others use legal prescription drugs incorrectly by taking too many (or too few) pills, using other people's medications, or self-medicating. Alcohol can also be abused by consuming to the point of intoxication. Other substances that are legal in some states (e.g., marijuana) can cause prolonged impairment with heavy usage.

It is essential that crew leaders enforce company policies and procedures regarding substance abuse. Crew leaders must work with management to deal with suspected drug and alcohol abuse and should not deal with these situations themselves. The Human Resources department or a designated manager usually handles these cases.

There are legal consequences of substance abuse and the associated safety implications. If you observe an employee showing impaired behavior for any reason, immediately contact your supervisor and/or Human Resources department for assistance. You protect the business, and the employee's and other workers' safety by taking these actions. It is the crew leader's responsibility to maintain safe working conditions at all times. This may include removing workers from a work site where they may be endangering themselves or others.

For example, suppose several crew members go out and smoke marijuana or drink during lunch. Then, they return to work to erect scaffolding for a concrete pour in the afternoon. If you can smell marijuana on the crew member's clothing or alcohol on their breath, you must step in and take action. Otherwise, they might cause an accident that could delay the project, or cause serious injury or death to themselves or others.

> *"Concern for man himself and his safety must always form the chief interest of all technical endeavors."*
>
> *—Albert Einstein*

It is often difficult to detect drug and alcohol abuse because the effects can be subtle. The best way is to look for identifiable effects, such as those mentioned above, or sudden changes in behavior that aren't typical of the employee. Some examples of such behaviors include the following:

- Unscheduled absences; failure to report to work on time
- Significant changes in the quality of work
- Unusual activity or **lethargy**
- Sudden and irrational temper flare-ups
- Significant changes in personal appearance, cleanliness, or health

There are other more specific signs that should arouse suspicion, especially if more than one is visible:

- Slurring of speech or an inability to communicate effectively
- Shiftiness or sneaky behavior, such as an employee disappearing to wooded areas, storage areas, or other private locations
- Wearing sunglasses indoors or on overcast days to hide dilated or constricted pupils, conditions which impair vision (*Figure 16*)

Figure 16 An impaired worker is a dangerous one.

- Wearing long-sleeved garments, particularly on hot days, to cover marks from needles used to inject drugs
- Attempting to borrow money from co-workers
- The loss of an employee's tools or company equipment

3.4.6 Job-Related Accident Investigations

Crew leaders are sometimes involved with an accident investigation. When an accident, injury, or report of work-connected illness takes place. If present on site, the crew leader should proceed immediately to the accident location to ensure that the victim receives proper first aid. The crew leader will also want to make sure that responsible individuals take other safety and operational measures to prevent another incident.

If required by company policy, the crew leader will also need to make a formal investigation and submit a report after an incident (including the completion of an *OSHA Form 301* report). An investigation looks for the causes of the accident by examining the circumstances under which it occurred and talking to the people involved. Investigations are perhaps the most useful tool in the prevention of future accidents.

The following are four major parts to an accident investigation:

- Describing the accident and related events leading to the accident
- Determining the cause(s) of the accident
- Identifying the persons or things involved and the part played by each
- Determining how to prevent reoccurrences

3.4.7 Promoting Safety

The best way for crew leaders to encourage safety is through example. Crew leaders should be aware that their behavior sets standards for their crew members. If a crew leader cuts corners on safety, then the crew members may think that it is okay to do so as well.

CAUTION

Workers often "follow the leader" when it comes to unsafe work practices. It is common for supervisory personnel to engage in unsafe practices and take more risks because they are more experienced. However, inexperienced or careless workers who take the same risks won't be as successful avoiding injury. As a leader, you must follow all safety practices to encourage your crew to do the same.

"I cannot trust a man to control others who cannot control himself."
—Robert E. Lee

The key to effectively promote safety is good communication. It is important to plan and coordinate activities and to follow through with safety programs. The most successful safety promotions occur when employees actively participate in planning and carrying out activities.

Some activities used by organizations to help motivate employees on safety and help promote safety awareness include the following:

- Safety training sessions
- Contests
- Recognition and awards
- Publicity

Safety training sessions can help keep workers focused on safety and give them the opportunity to discuss safety concerns with the crew. A previous section addressed this topic.

Did You Know?

Substance Abuse

An employee who is involved in an accident while under the influence of drugs or alcohol may be denied workers compensation insurance benefits.

Case Study

For years, a prominent safety engineer was confused as to why sheet-metal workers fractured their toes frequently. The crew leader had not performed thorough accident investigations, and the injured workers were embarrassed to admit how the accidents really occurred. Further investigation discovered they used the metal-reinforced cap on their safety shoes as a "third hand" to hold the sheet metal vertically in place when they fastened it. The rigid and heavy metal sheet was inclined to slip and fall behind the safety cap onto the toes, causing fractures. The crew leader could have prevented several injuries by performing a proper investigation after the first accident.

3.4.8 Safety Contests

Contests are a great way to promote safety in the workplace. Examples of safety-related contests include the following:

- Sponsoring housekeeping contests for the cleanest job site or work area
- Challenging employees to come up with a safety slogan for the company or department
- Having a poster contest that involves employees or their children creating safety-related posters
- Recording the number of accident-free workdays or worker-hours
- Giving safety awards (hats, T-shirts, other promotional items or prizes)

One of the positive aspects of safety contests is their ability to encourage employee participation. It is important, however, to ensure that the contest has a valid purpose. For example, workers can display the posters or slogans created in a poster contest throughout the organization as safety reminders.

CAUTION

One mistake that some companies make when offering safety contests is providing tangible or monetary awards to departments or teams specifically for the lowest number of reported accidents or accident-free work hours. While well-intentioned, this approach appeals to the tendency of people to inflate their performance to win. Consequently, history shows this has the negative effect of encouraging the under-reporting of accidents and injuries, which defeats the purposes of safety contests.

3.4.9 Incentives and Rewards

Incentives and awards serve several purposes. Among them are acknowledging and encouraging good performance, building goodwill, reminding employees of safety issues, and publicizing the importance of practicing safety standards. There are countless ways to recognize and award safety. Examples include the following:

- Supplying food at the job site when a certain goal is achieved

- Providing a reserved parking space to acknowledge someone for a special achievement
- Giving gift items such as T-shirts or gift certificates to reward employees
- Giving awards to a department or an individual (*Figure 17*)
- Sending a letter of appreciation
- Publicly honoring a department or an individual for a job well done

You can use creativity to determine how to recognize and award good safety on the work site. The only precautionary measure is that the award should be meaningful and not perceived as a bribe. It should be representative of the accomplishment.

3.4.10 Publicity

Publicizing safety is the best way to get the message out to employees. An important aspect of publicity is to keep the message accurate and current. Safety posters that hang for years on end tend to lose effectiveness. It is important to keep ideas fresh.

Examples of promotional activities include posters or banners, advertisements or information on bulletin boards, payroll mailing stuffers, and employee newsletters. In addition, the company can purchase merchandise that promotes safety, including buttons, hats, T-shirts, and mugs.

Figure 17 An example of a safety award.

Participant Exercise D

Described here are three scenarios that reflect unsafe practices by craft workers. For each of these scenarios write down how you would deal with the situation, first as the crew leader of the craft worker, and then as the leader of another crew.

1. You observe a worker wearing his hard hat backwards and his safety glasses hanging around his neck. He is using a concrete saw.

2. As you are supervising your crew on the roof deck of a building under construction, you notice that a section of guard rail has been removed. Another contractor was responsible for installing the guard rail.

3. Your crew is part of plant shutdown at a power station. You observe that a worker is welding without a welding screen in an area where there are other workers.

Additional Resources

Construction Workforce Development Professional, NCCER. 2016. New York, NY: Pearson Education, Inc.

Mentoring for Craft Professionals, NCCER. 2016. New York, NY: Pearson Education, Inc.

The following websites offer resources for products and training:

National Census of Fatal Occupational Injuries (NCFOI), **www.bls.gov**

National Institute of Occupational Safety and Health (NIOSH), **www.cdc.gov/niosh**

National Safety Council, **www.nsc.org**.

Occupational Safety and Health Administration (OSHA), **www.osha.gov**

3.0.0 Section Review

1. One of a crew leader's most important responsibilities to the employer is to _____.
   a. enforce company safety policies
   b. estimate material costs for a project
   c. make recommendations for setting up a crew
   d. provide input for fixed-price contracts

2. What amount can OSHA fine a company for willfully committing a violation of an OSHA safety standard?
   a. $1,000
   b. $7,000
   c. $12,471
   d. $124,709

3. Who is ultimately responsible for a worker's safety?
   a. The individual worker
   b. The crew leader
   c. The project superintendent
   d. The company HR department head

4. A crew leader's safety responsibilities include all of the following *except* _____.
   a. conducting safety training sessions
   b. developing a company safety program
   c. performing safety inspections
   d. participating in accident investigations

Section Four

4.0.0 Project Planning

Objective

Demonstrate a basic understanding of the planning process, scheduling, and cost and resource control.

a. Describe how construction contracts are structured.
b. Describe the project planning and scheduling processes.
c. Explain how to implement cost controls on a construction project.
d. Explain the crew leader's role in controlling project resources and productivity.

Performance Tasks

1. Develop and present a look-ahead schedule.
2. Develop an estimate for a given work activity.

Trade Terms

Critical path: In manufacturing, construction, and other types of creative processes, the required sequence of tasks that directly controls the ultimate completion date of the project.

Job diary: A written record that a supervisor maintains periodically (usually daily) of the events, communications, observations, and decisions made during the course of a project.

Look-ahead schedule: A manual- or software-scheduling tool that looks several weeks into the future at the planned project events; used for anticipating material, labor, tool, and other resource requirements, as well as identifying potential schedule conflicts or other problems.

Return on investment (ROI): A measure of the gain or loss of money resulting from an investment; normally measured as a percentage of the original investment.

Work breakdown structure (WBS): A diagrammatic and conceptual method for subdividing a complex project, concept, or other thing into its various functional and organizational parts so that planners can analyze and plan for each part in detail.

This section describes methods of efficient project control. It examines estimating, planning and scheduling, and resource and cost control. All workers who participate in a job are responsible at some level for controlling cost and schedule performance, and for ensuring that they complete the project according to plans and specifications.

> **NOTE**
> This section mainly pertains to building-construction projects, but the project control principles described here apply generally to all types of projects.

The contractor, project manager, superintendent, and crew leader each have management responsibilities for their assigned jobs. For example, the contractor's responsibility begins with obtaining the contract, and it doesn't end until the client takes ownership of the project. The project manager is generally the person with overall responsibility for coordinating the project. Finally, the superintendent and crew leader are responsible for coordinating the work of one or more workers, one or more crews of workers within the company or, on occasion, one or more crews of subcontractors. The crew leader directs a crew in the performance of work tasks.

4.1.0 Construction Project Phases, Contracts, and Budgeting

Construction projects consist of three phases: the *development phase*, the *planning phase*, and the *construction phase*. Throughout these phases, the property owner must work directly with engineers and architects to gather data necessary for persuading a contractor to accept the job. Once a job is in progress, the crew leader fills an active role in maintaining the planned budget for the project.

4.1.1 Development Phase

A new building project begins when an owner has decided to build a new facility or add to an existing facility. The development process is the first stage of planning for a new building project. This process involves land research and feasibility studies to ensure that the project has merit. Architects or engineers develop the conceptual drawings that define the project graphically. They then provide the owner with sketches of room layouts and elevations and make suggestions about what construction materials to use.

During the development phase, architects, engineers, and/or the owner develop an estimate for the proposed project and establish a preliminary budget. Once that budget is established, the financing of the project with lending institutions begins. The development team begins preliminary reviews with government agencies. These reviews include zoning, building restrictions, landscape requirements, and environmental impact studies.

The owner must analyze the project's cost and potential **return on investment (ROI)** to ensure that its costs won't exceed its market value and that the project provides a reasonable profit during its existence. If the project passes this test, the responsible architects/engineers will proceed to the planning phase.

4.1.2 Planning Phase

When the architects/engineers begin to develop the project drawings and specifications, they consult with other design professionals such as structural, mechanical, and electrical engineers. They perform the calculations, make a detailed technical analysis, and check details of the project for accuracy.

The design professionals create drawings and specifications. They use these drawings and specifications to communicate the necessary information to the contractors, subcontractors, suppliers, and workers that contribute to a project.

During the planning phase, the owners hold many meetings (*Figure 18*) to refine estimates, adjust plans to conform to regulations, and secure a construction loan. If the project is a condominium, an office building, or a shopping center, then a marketing firm develops a marketing program. In such cases, the selling of the project often starts before actual construction begins.

Figure 18 Architects and clients meet to refine plans.

Next, the design team produce a complete set of drawings, specifications, and bid documents. Then the owner will select the method to obtain contractors. The owner may choose to negotiate with several contractors or select one through competitive bidding. Everyone concerned must also consider safety as part of the planning process. A safety crew leader may walk through the site as part of the pre-bid process.

Contracts for construction projects can take many forms. All types of contracts fall under three basic categories: *firm-fixed-price*, *cost reimbursable*, and *guaranteed maximum price*.

Firm-fixed-price – In this type of contract, the buyer generally provides detailed drawings and specifications, which the contractor uses to calculate the cost of materials and labor. To these costs, the contractor adds a percentage representing company overhead expenses such as office rent, insurance, and accounting/payroll costs. At the end, the contractor adds a profit factor.

When submitting the bid, the contractor will state very specifically the conditions and assumptions on which the company based the bid. These conditions and assumptions also form the basis from which parties can price allowable changes to the contract. Because contracting parties establish the price in advance, any changes in the job requirements once the job is started will impact the contractor's profit margin.

This is where the crew leader can play an important role by identifying problems that increase the amount of planned labor or materials. By passing this information up the chain of command, the crew leader allows the company to determine if the change is outside the scope of the bid. If so, they can submit a change order request to cover the added cost.

Cost reimbursable contract – In this type of contract, the buyer reimburses the contractor for labor, materials, and other costs encountered in the performance of the contract. Typically, the contractor and buyer agree in advance on hourly or daily labor rates for different categories of worker. These rates include an amount representing the contractor's overhead expense. The buyer also reimburses the contractor for the cost of materials and equipment used on the job.

The buyer and contractor also negotiate a profit margin. On this type of contract, the profit margin is likely to be lower than that of a fixed-price contract because of the significantly-reduced contractor's cost risk. The profit margin is often subject to incentive or penalty clauses that make the amount of profit awarded subject to performance by the contractor. The contract usually ties performance to project schedule milestones.

Guaranteed maximum price (GMP) contract – This form of contract, also called a *not-to-exceed contract*, is common on projects negotiated mainly with the owner. Owner's involvement in the process usually includes preconstruction, and the entire team develops the parameters that define the basis for the work.

With a GMP contract, the owner reimburses the contractor for the actual costs incurred by the contractor. The contract also includes a payment of a fixed fee up to the maximum price allowed in the contract. The contractor bears any cost overruns.

The advantages of the GMP contract vehicle may include the following:

- Reduced design time
- Allows for phased construction
- Uses a team approach to a project
- Reduction in changes related to incomplete drawings

4.1.3 Construction Phase

The designated contractor enlists the help of mechanical, electrical, elevator, and other specialty subcontractors to complete the construction phase. The contractor may perform one or more parts of the construction, and rely on subcontractors for the remainder of the work. However, the general contractor is responsible for managing all the trades necessary to complete the project. *Figure 19* shows the flow of a typical project from beginning to end.

As construction nears completion, the architect/engineer, owner, and government agencies start their final inspections and acceptance of the project. If the general contractor has managed the project, the subcontractors have performed their work, and the architects/engineers have regularly inspected the project to ensure it satisfied the local code, then the inspection process can finish up in a timely manner. This results in a satisfied client and a profitable project for all.

On the other hand, if the inspection reveals faulty workmanship, poor design, incorrect use of materials, or violation of codes, then the inspection and acceptance will become a lengthy battle and may result in a dissatisfied client and an unprofitable project.

The initial set of drawings for a construction project reflects the completed project as conceived by the architect and engineers. During construction, changes are usually necessary because of factors unforeseen during the design phase. For example, when electricians must reroute cabling or conduit, or the installed equipment location is different than shown on the original drawing, such changes must be marked on the drawings. Without this record, technicians called to perform maintenance or modify the equipment later will have trouble locating all the cabling and equipment.

Project supervision must document any changes made during construction or installation on the drawings as the changes occur. Architects usually note changes on hard-copy drawings using a colored pen or pencil, so users can readily spot the change. These marked-up versions are commonly called **redline drawings**. With mobile digital technology, architects and engineers can revise and promulgate the latest drawing versions almost instantly. After the drawings have been revised to reflect the redline changes, the final drawings are called **as-built drawings**, and are so marked. These become the drawings of record for the project.

4.1.4 Project-Delivery Systems

Project-delivery systems are processes for constructing projects, from development through construction. Project delivery systems focus on the following three primary systems, shown in *Figure 20*:

- *General contracting* – The traditional project delivery system uses a general contractor. In this type of project, the owner determines the design of the project, and then solicits proposals from general contractors. After selecting a general contractor, the owner contracts directly with that contractor, who builds the project as the prime, or controlling, contractor.
- *Design-build* – In the design-build system, a single entity manages both the design and construction of a project. Design-build delivery commonly use GMP contracts.
- *Construction management* – The construction management project delivery system uses a construction manager to facilitate the design and construction of a project. Construction managers are very involved in project control; their main concerns are controlling time, cost, and the quality of the project.

4.1.5 Cost Estimating and Budgeting

Before building a project, an estimate must be prepared. Estimating is the process of calculating the cost of a project. There are two types of costs to consider, including direct and indirect costs. Direct costs, also known as general conditions, are those that planners can clearly assigned to a

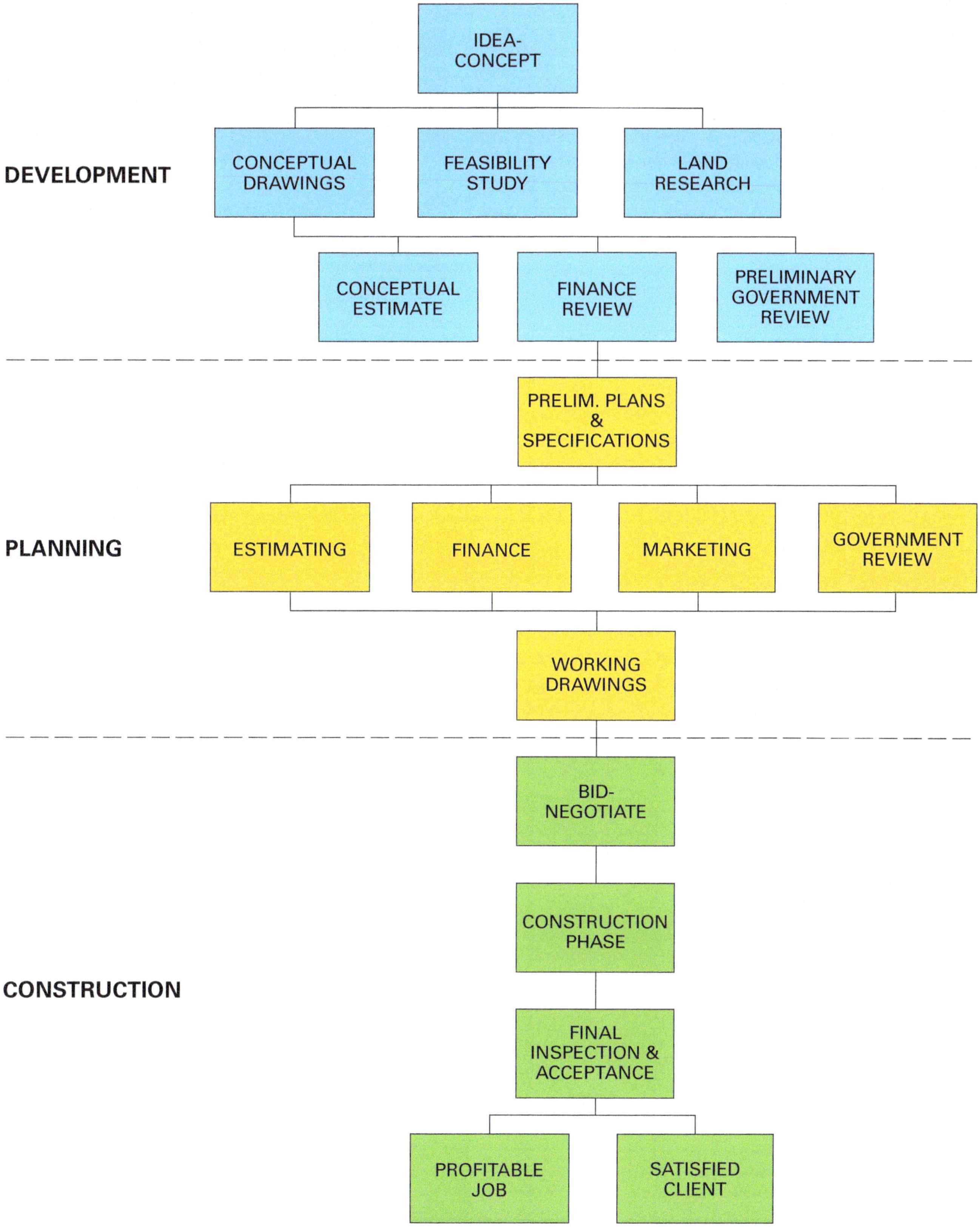

Figure 19 Project flow diagram.

GOING GREEN

LEED

Figure Credit: © Linda Williams/Dreamstime.com

LEED stands for *Leadership in Energy and Environmental Design*, which is an initiative started by the US Green Building Council (USGBC) to encourage and accelerate the adoption of sustainable construction standards worldwide through its Green Building Rating System. USGBC is a non-government, not-for-profit group. Their rating system addresses six categories:

1. Sustainable Sites (SS)
2. Water Efficiency (WE)
3. Energy and Atmosphere (EA)
4. Materials and Resources (MR)
5. Indoor Environmental Quality (EQ)
6. Innovation in Design (ID)

Building owners are the driving force behind the LEED voluntary program. Construction crew leaders may not have input into the decision to seek LEED certification for a project, or what materials are used in the project's construction. However, crew leaders can help to minimize material waste and support recycling efforts, both of which are factors in obtaining LEED certification.

An important question to ask is whether your project is seeking LEED certification. If the project is seeking certification, the next step is to ask what your role will be in getting the certification. If you are procuring materials, what information does the project need and who should receive it? What specifications and requirements do the materials need to meet? If you are working outside the building or inside in a protected area, what do you need to do to protect the work area? How does your crew manage waste? Are there any other special requirements that will be your responsibility? Do you see any opportunities for improvement? LEED principles are described in more detail in NCCER's *Your Role in the Green Environment* (Module ID 70101-15) and *Sustainable Construction Supervisor* (Module ID 70201-11).

DELIVERY SYSTEMS

RESPONSIBLE PARTIES	GENERAL CONTRACTING	DESIGN-BUILD	CONSTRUCTION MANAGEMENT
OWNER	Designs project (or hires architect)	Hires general contractor	Hires construction management company
GENERAL CONTRACTOR	Builds project (with owner's design)	Involved in project design, builds project	Builds, may design (hired by construction management company)
CONSTRUCTION MANAGEMENT COMPANY			Hires and manages general contractor and architect

Figure 20 Project-delivery systems.

budget. Indirect costs are overhead costs shared by all projects. Planners generally calculate these costs as an overhead percentage to labor and material costs.

Direct costs include the following:

- Materials
- Labor
- Tools
- Equipment

Indirect costs refer to overhead items such as the following:

- Office rent
- Utilities
- Telecommunications
- Accounting
- Office supplies, signs

The bid price includes the estimated cost of the project as well as the profit. Profit refers to the amount of money that the contractor will make after paying all the direct and indirect costs. If the direct and indirect costs exceed the estimate for the job, the difference between the actual and estimated costs must come out of the company's profit. This reduces what the contractor makes on the job.

Profit is the fuel that powers a business. It allows the business to invest in new equipment and facilities, provide training, and to maintain a reserve fund for times when business is slow. In large companies, profitability attracts investors who provide the capital necessary for the business to grow. For these reasons, contractors can't afford to consistently lose money on projects. If they can't operate profitably, they are forced out of business. Crew leaders can help their companies remain profitable by managing budget, schedule, quality, and safety adhering to the drawings, specifications, and project schedule.

The cost estimate must consider many factors. Many companies employ professional cost estimators to do this work. They also maintain performance data for previous projects. They use this data as a guide in estimating new projects. Development of a complete estimate generally proceeds as follows:

Step 1 Using the drawings and specifications, an estimator records the quantity of the materials needed to construct the job. Construction companies call this step the *quantity takeoff*. The estimator enters the information on a hard-copy or digital takeoff sheet like the one shown in *Figure 21*.

Step 2 The estimator uses the company's productivity rates to calculate the amount of labor required to complete the project. Most companies keep records of these rates for the type and size of the jobs that they perform. The company's estimating department maintains and updates these records.

Step 3 The estimator calculates the labor hours required by dividing the estimated amount of work by the productivity rate.

- For example, if the productivity rate for concrete finishing is 40 square feet per hour, and there are 10,000 square feet of concrete to be finished, then 250 hours of concrete finishing labor is required (10,000 ÷ 4 = 250).
- The estimator multiplies this number by the hourly rate for concrete finishing to determine the cost of that labor category.
- If this work is subcontracted, then the estimator uses the subcontractor's cost estimate, raised by an overhead factor, in place of direct-labor cost.

Step 4 The estimator transfers the total material quantities from the quantity takeoff sheet to a summary or pricing sheet (*Figure 22*). The total cost of materials is calculated after obtaining material prices from local suppliers.

Step 5 Next, the estimator determines the cost of equipment needed for the project. This number could reflect rental cost or a factor applied by the company when they plan to use their own equipment.

Step 6 The estimator totals the cost of all resources on the summary sheet—materials, equipment, tools, and labor. The estimator can also calculate the material unit cost—the total cost divided by the total number of units of listed materials.

Step 7 The estimator adds the cost of taxes, bonds, insurance, subcontractor work, and other indirect costs to the direct costs of the materials, equipment, tools, and labor.

Step 8 A sum of direct and indirect costs yields the total project cost. The contractor adds the expected profit to that total.

WORKSHEET

Takeoff By: | DATE ______ PROJECT ______

Checked By: | SHEET ___ of ___ ARCHITECT ______

PAGE #

REF.	DESCRIPTION	DIMENSIONS				EXTENSION	QUANTITY	UNIT	TOTAL		REMARKS
		NO	LENGTH	WIDTH	HEIGHT				QUANTITY	UNIT	

Figure 21 Quantity takeoff sheet.

SUMMARY SHEET

By:

PAGE #

DATE ______ SHEET ______ of ______

TITLE: ______

PROJECT ______ WORK ORDER # ______

	DESCRIPTION	QUANTITY		MATERIAL COST		LABOR MAN HOURS FACTORS					LABOR COST		ITEM COST	
		TOTAL	UT	PER UNIT	TOTAL	CRAFT	PR UNIT	TOTAL	RATE	COST PR	PER	TOTAL	TOTAL	PER UNIT
					MATERIAL							LABOR	TOTAL	

Figure 22 Summary sheet.

> **NOTE**
>
> There are computer programs and mobile applications available to simplify the cost-estimating process. Many of them are tailored to specific industries, such as construction or manufacturing, as well as specific trades within the industries, such as plumbing, electrical, HVAC, and insulating contractors.
>
> Estimating programs are typically set up to include a takeoff form and a form for estimating labor by category. Most of these programs include a database that contains current prices for labor and materials, so they automatically price the job and produce a bid. The programs can also generate purchase orders for materials on the job.

As a crew leader, you may be required to estimate quantities of materials. You will need a set of construction drawings and specifications to estimate the amount of a certain type of material required to perform a job. You should carefully review the appropriate section of the technical specifications and page(s) of drawings to determine the types and quantities of materials required. Then enter the quantities on the worksheet. For example, you should review the specification section on finished carpentry along with the appropriate pages of drawings before taking off the linear feet of door and window trim.

If insufficient materials are available to complete the job and an estimate is required, the estimator must determine how much more construction work is necessary. Knowing this, the crew leader can then determine the materials needed. You must also reference the construction drawings in this process.

4.2.0 Planning

One definition of planning is determining the methods used and their sequence to carry out the different tasks to complete a project (*Figure 23*). Planning involves the following:

- Determining the best method for performing the job
- Identifying the responsibilities of each person on the work crew
- Determining the duration and sequence of each activity
- Identifying the tools and equipment needed to complete a job
- Ensuring that the required materials are at the work site when needed
- Making sure that heavy construction equipment is available when required

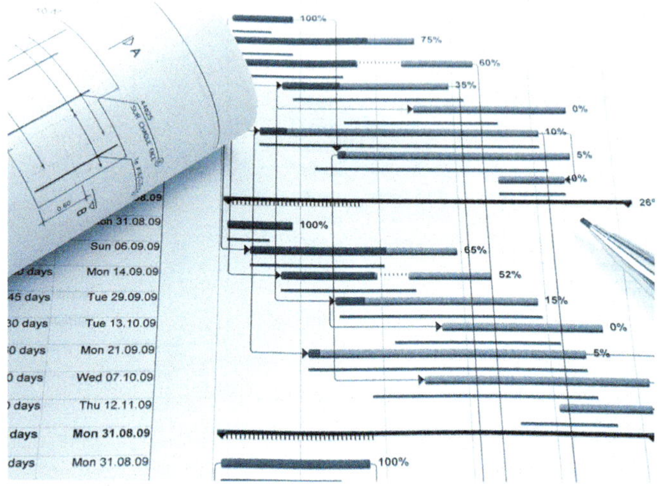

Figure 23 Proper prior planning prevents poor performance.

- Working with other contractors in such a way as to avoid interruptions and delays

Some important reasons for crew-leader planning include the following:

- Controlling the job in a safe manner so that it is built on time and within cost
- Lowering job costs through improved productivity
- Preparing for bad weather or unexpected occurrences
- Promoting and maintaining favorable employee morale
- Determining the best and safest methods for performing the job

With a plan, a crew leader can direct work efforts efficiently and can use resources such as personnel, materials, tools, equipment, work area, and work methods to their full potential.

A proactive crew leader will always have a backup plan in case circumstances prevent the original plan from working. There are many circumstances that can cause a plan to go awry, including adverse weather, equipment failure, absent crew members, and schedule slippage by other crafts.

Project planners establish time and cost limits for the project; the crew leader's planning must fit within those constraints. Therefore, it is important to consider the following factors that may affect the outcome:

- Site and local conditions, such as soil types, accessibility, or available staging areas
- Climate conditions that should be anticipated during the project
- Timing of all phases of work
- Types of materials to be installed and their availability

Participant Exercise E

Assume you are the leader of a crew building footing formwork for the construction shown in the figure below. You have used all of the materials provided for the job, yet you have not completed it. You study the drawings and see that the formwork consists of two side forms, each 12" high. The total length of footing for the entire project is 115'-0". You have completed 88'-0" to date; therefore, you have 27'-0" remaining (115' – 88' = 27').

Your task is to prepare an estimate of materials that you will need to complete the job. In this case, you will estimate only the side form materials (do not consider the miscellaneous materials here).

- Footing length to complete: 27'-0"
- Footing height: 1'-0"

Refer to the worksheet on the next page for a final tabulation of the side forms needed to complete the job.

1. Using the same footing as described in the example above, calculate the quantity (square feet) of formwork needed to finish 203 linear feet of the footing. Place this information directly on the worksheet.
2. You are the crew leader of a carpentry crew whose task is to side a warehouse with plywood sheathing. The wall height is 16', and there is a total of 480 linear feet of wall to side. You have done 360 linear feet of wall and have run out of materials. Calculate how many more feet of plywood you will need to complete the job. If you are using 4 × 8 plywood panels, how many will you need to order to cover the additional work? Write your estimate on the worksheet.

Show your calculations to the instructor.

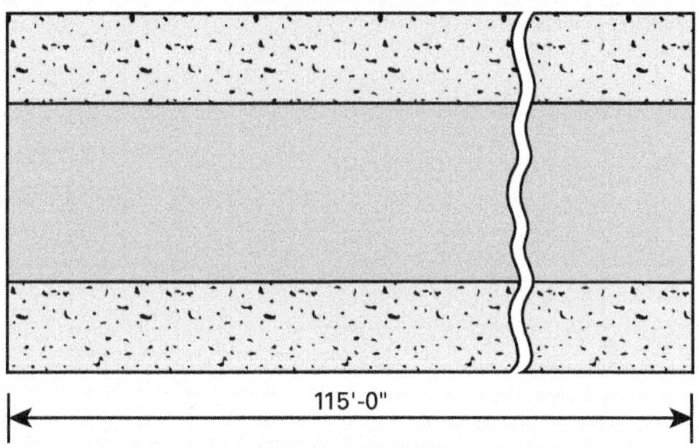

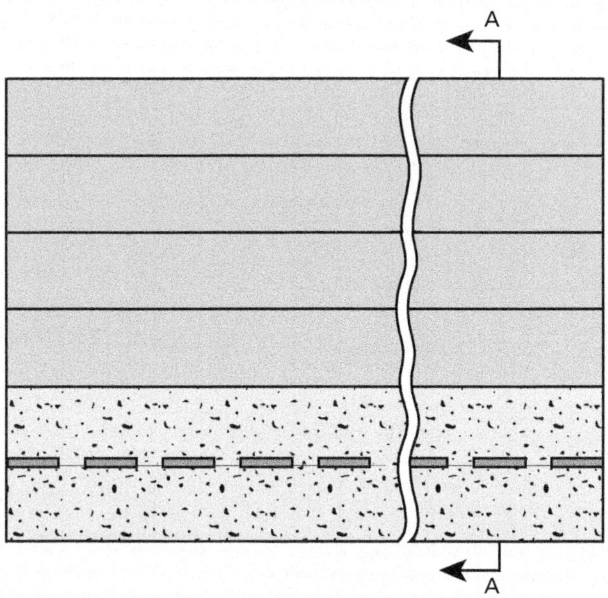

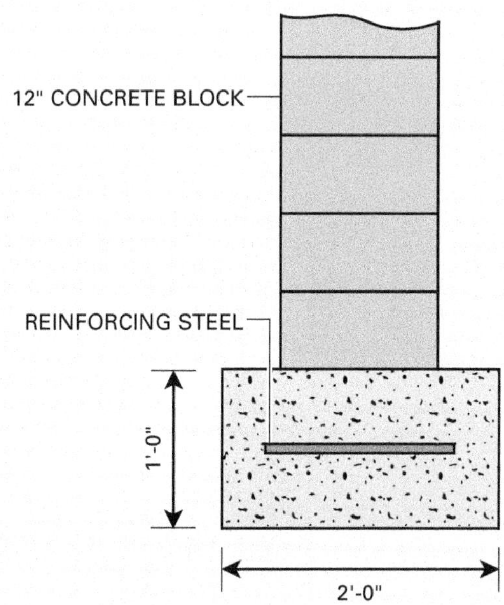

WORKSHEET

Takeoff By: RWH	DATE 2/1/17	PROJECT Sam's Diner	PAGE #1
Checked By:	SHEET 01 of 01	ARCHITECT 654b	

REF.	DESCRIPTION	DIMENSIONS				EXTENSION	QUANTITY	UNIT	TOTAL		REMARKS
		NO	LENGTH	WIDTH	HEIGHT				QUANTITY	UNIT	
	Footing Side Forms 2		27'0"		1'0"	2x27x1	54	SF	54	SF	

- Equipment and tools required and their availability
- Personnel requirements and availability
- Relationships with the other contractors and their representatives on the job

On a simple job, crews can handle these items almost automatically. However, larger or more complex jobs require the planner to give these factors more formal consideration and study.

4.2.1 Stages of Planning

Formal planning for a construction job occurs at specific times in the project process. The two most important stages of planning occur in the preconstruction phase and during the construction work.

Preconstruction planning – The preconstruction stage of planning occurs before the start of construction. The preconstruction planning process doesn't always involve the crew leader, but it's important to understand what it consists of.

There are two phases of preconstruction planning. The first is developing the proposal, bid, or negotiated price for the job. This is when the estimator, the project manager, and the field superintendent develop a preliminary plan for completing the work. They apply their experience and knowledge from previous projects to develop the plan. The process involves determining what methods, personnel, tools, and equipment the work will require and what level of productivity they can achieve.

The second phase of preconstruction planning occurs after the client awards the contract. This phase requires a thorough knowledge of all project drawings and specifications. During this process, planners select the actual work methods and resources needed to perform the work. Here, crew leaders might get involved, but their planning must adhere to work methods, production rates, and resources that fit within the estimate prepared during contract negotiations. If the project requires a method of construction different

Participant Exercise F

1. In your own words, define planning, and describe how a job can be done better if it is planned. Give an example.

____

____

____

____

2. Consider a job that you recently worked on to answer the following:
   a. List the material(s) used.
   b. List each member of the crew with whom you worked and what each person did.
   c. List the kinds of equipment used.

____

____

____

____

____

3. List some suggestions for how the job could have been done better, and describe how you would plan for each of the suggestions.

____

____

____

____

from what is normal, planners will usually inform the crew leader of what method to use.

Construction planning – During construction, the crew leader is directly involved in daily planning. This stage of planning consists of selecting methods for completing tasks before beginning work. Effective planning exposes likely difficulties, and enables the crew leader to minimize the unproductive use of personnel and equipment. Proper planning also provides a gauge to measure job progress.

Effective crew leaders develop a tool known as the **look-ahead schedule**. These schedules consider actual circumstances as well as projections two-to-three weeks into the future. Developing a look-ahead schedule helps ensure that all resources are available on the project when needed. Most scheduling apps and programs include this feature that automatically help you focus on this period in the job.

4.2.2 The Planning Process

The planning process consists of the following five steps:

Step 1 Establish a goal.

Step 2 Identify the completed work activities required to achieve the goal.

Step 3 Identify the required tasks to accomplish those activities.

Step 4 Communicate responsibilities.

Step 5 Follow up to verify the goal achievement.

Establishing a goal – The term *goal* has different meanings for different people. In general, a goal is a specific outcome that one works toward. For example, the project superintendent of a home construction project could establish the goal to have a house dried-in by a certain date. (The term *dried-in* means ready for the application of roofing and siding.) To meet that goal, the leader of the framing crew and the superintendent would need to agree to a goal to have the framing completed by a given date. The crew leader would then establish sub-goals (objectives) for the crew to complete each element of the framing (floors, walls, roof) by a set time. The superintendent would need to set similar goals with the crews that install sheathing, building wrap, windows, and exterior doors. However, if the framing crew doesn't meet its goal, that will delay the other crews.

Identifying the required work – The second step in planning is to identify the necessary work to achieve the goal as a series of activities in a certain sequence. You will learn how to break down a job into activities later in this section. At this point, the crew leader should know that, for each activity, one or more objectives must be set.

An objective is a statement of a condition the plan requires to exist or occur at a specific time. An objective must:

- Mean the same thing to everyone involved
- Be measurable, so that everyone knows when it has been reached
- Be achievable with the resources available
- Have everyone's full support

Examples of practical objectives include the following:

- "By 4:30 p.m. today, the crew will have completed installation of the floor joists."
- "By closing time Friday, the roof framing will be complete."

Notice that both examples meet the first three requirements of an objective. Planners assume that everyone involved in completing the task is committed to achieving the objective. The advantage in developing objectives for each work activity is that it allows the crew leader to determine if the crew is following the plan. In addition, objectives serve as sub-goals that are usually under the crew leader's control.

Some construction work activities, such as installing 12" footing forms, are done so often that they require little planning. However, other jobs, such as placing a new type of mechanical equipment, require substantial planning. This type of job demands that the crew leader set specific objectives.

Whenever faced with a new or complex activity, take the time to establish objectives that will serve as guides for accomplishing the job. You can use these guides in the current situation, as well as for similar work in the future.

Identifying the required tasks – To plan effectively, the crew leader must be able to break a work activity assignment down into smaller tasks. Large jobs include a greater number of tasks than small ones, but all jobs can be broken down into manageable components.

When breaking down an assignment into tasks, make each task identifiable and definable. A task is identifiable when one knows the types and amounts of resources it requires. A task is definable if it has a specific duration. For purposes of efficiency, the job breakdown should not be too detailed or complex, unless the job has never been done before or must be performed with strictest efficiency.

For example, a suitable breakdown for the work activity to install square vinyl floor tiles in a cafeteria might be the following:

Step 1 Prepare the floor.

Step 2 Stage the tiles.

Step 3 Spread the adhesive.

Step 4 Lay the tiles.

Step 5 Clean the tiles.

Step 6 Wax the floor.

The crew leader could create even more detail by breaking down any one of the tasks into subtasks. In this case, however, that much detail is unnecessary and wastes the crew leader's time and the project's money.

Planners can divide every work activity into three general parts:

- Preparing
- Performing
- Cleaning up

Some of the most frequent mistakes made in the planning process are forgetting to prepare and to clean up. The crew leader must not overlook preparation and cleanup.

After identifying the various tasks that make up the job and developing an objective for each task, the crew leader must determine what resources the job requires. Resources include labor, equipment, materials, and tools. In most jobs, the job estimate identifies these resources. The crew leader must make sure that these resources are available on the site when needed.

"By failing to prepare, you are preparing to fail."
—Benjamin Franklin

Communicating responsibilities – No supervisor can complete all the activities within a job alone. The crew leader must rely on other people to get everything done. Therefore, most jobs have a crew of people with various experiences and skill levels to assist in the work. The crew leader's job is to draw from this expertise to get the job done well and in a safe and timely manner.

Once the various activities that make up the job have been determined, the crew leader must identify the person or persons responsible for completing each activity. This requires that the crew leader be aware of the skills and abilities of the people on the crew. Then, the crew leader must put this knowledge to work in matching the crew's skills and abilities to specific tasks required to complete the job.

After matching crew members to specific activities, the crew leader must then assign work to the crew. The crew leader normally communicates responsibilities verbally; the crew leader often talks directly to the responsible person for the activity. There may be times when the crew leader assigns work through written instructions or indirectly through someone other than the crew leader. Either way, crew members should know what it is they are responsible for accomplishing on the job.

Following up – Once the crew leader has delegated the activities to the appropriate crew members, there must be follow-up to make sure that the crew has completed them correctly and efficiently. Task follow-up involves being present on the job site to make sure all the resources are available to complete the work, ensuring that the crew members are working on their assigned activities, answering any questions, and helping to resolve any problems that occur while the work is being done. In short, follow-up activity means that the crew leader is aware of what's going on at the job site and is doing whatever is necessary to make sure that the crew completes the work on schedule. *Figure 24* reviews the planning steps.

The crew leader should carry a small note pad or electronic device for planning and taking notes. That way, you can record thoughts about the project as they occur, and you won't forget pertinent details. The crew leader may also choose to use a manual planning form such as the one illustrated in *Figure 25*.

As the job progresses, refer to these resources to see that the tasks are being done according to plan. This is job analysis. Construction projects that don't proceed according to work plans usually end up costing more and taking longer. Therefore, it is important that crew leaders refer to the planning documents periodically.

The crew leader is involved with many activities on a day-to-day basis. Thus, it is easy to forget important events if they aren't recorded. To help keep track of events such as job changes, interruptions, and visits, the crew leader should keep a **job diary**. A job diary is a notebook in which the crew leader records activities or events that take place on the

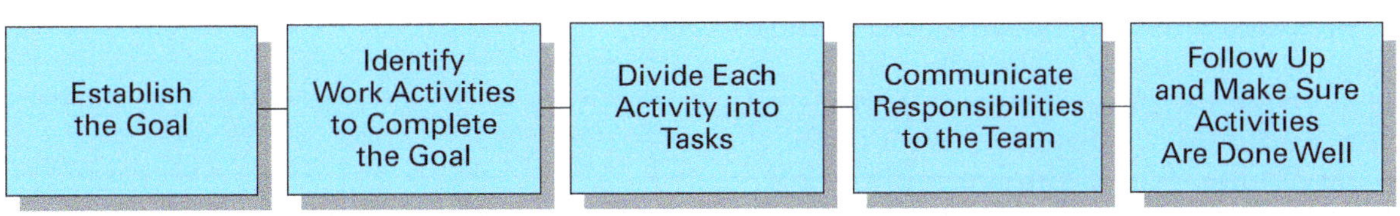

Figure 24 Steps to effective planning.

DAILY WORK PLAN

"PLAN YOUR WORK AND WORK YOUR PLAN = EFFICIENCY"

Plan of ____________ Date ________

PRIORITY	DESCRIPTION	✓ When Completed X Carried Forward

Figure 25 Planning form.

job site that may be important later. When making entries in a job diary, make sure that the information is accurate, factual, complete, consistent, organized, and up-to-date. This is especially true if documenting personnel problems. If the company requires maintaining a job diary, follow company policy in determining which events and what details you should record. However, if there is a doubt about what to include, it is better to have too much information than too little. *Figure 26* shows a sample page from a job diary.

4.2.3 Planning Resources

Once a job has been broken down into its tasks or activities, the various resources needed to perform them must be determined and accounted for. Resource planning includes the following specific considerations:

- *Safety planning* – Using the company safety manual as a guide, the crew leader must assess the safety issues associated with the job and take necessary measures to minimize any risk to the crew. This may involve working with the company or site safety officer and may require a formal hazard analysis.
- *Materials planning* – Preconstruction planning identifies the materials required for the job and lists them on the job estimate. Companies usually order the materials from suppliers who have previously provided quality materials on schedule and within estimated cost.

  The crew leader is usually not involved in the planning and selection of materials, which happens during the preconstruction phase. The crew leader does, however, have a major role to play in the receipt, storage, and control of the materials after they reach the job site.

- The crew leader is also involved in planning materials for tasks such as job-built formwork and scaffolding. In addition, the crew leader may run out of a specific material, such as fasteners, and need to order more. In such cases, be sure to consult the appropriate supervisor, since most companies have specific purchasing policies and procedures.

July 8, 2017

Weather: Hot and Humid

Project: Company XYZ Building

- The paving contractor crew arrived late (10 am).
- The owner representative inspected the footing foundation at approximately 1 pm.
- The concrete slump test did not pass. Two trucks had to be ordered to return to the plant, causing a delay.
- John Smith had an accident on the second floor. I sent him to the doctor for medical treatment. The cause of the accident is being investigated.

Figure 26 Sample page from a job diary.

- *Site planning* – There are many planning elements involved in site work. The following are some of the key elements:
  - Access roads
  - Emergency procedures
  - Material and equipment storage
  - Material staging
  - Parking
  - Sedimentation control
  - Site security
  - Storm water runoff
- *Equipment planning* – The preconstruction phase addresses much of the planning for use of construction equipment. This planning includes the types of equipment needed, the use of the equipment, and the length of time it will be on the site. The crew leader must work with the main office to make certain that the equipment reaches the job site on time. The crew leader must also ensure that crew equipment operators are properly trained.

  Coordinating the use of the equipment is also very important. Some equipment operates in combination with other equipment. For example, dump trucks are generally required when loaders and excavators are used. The crew leader should also coordinate equipment with other contractors on the job. Sharing equipment can save time and money and avoid duplication of effort.

  The crew leader must reserve time for equipment maintenance to prevent equipment failure. In the event of an equipment failure, the crew leader must know who to contact to resolve the problem. An alternate plan must be ready in case one piece of equipment breaks down, so that the other equipment doesn't sit idle. The crew leader should coordinate these contingency plans with the main office or the crew leader's immediate superior.
- *Tool planning* – A crew leader is responsible for planning tool usage for a job. This task includes the following:
  - Determining the tools required
  - Informing the workers who will provide the tools (company or worker)
  - Making sure the workers are qualified to use the tools safely and effectively
  - Determining what controls to establish for tools
- *Labor planning* – All jobs require some sort of labor because the crew leader can't complete all the work alone. When planning for labor, the crew leader must do the following:
  - Identify the skills needed to perform the work.
  - Determine the number of people having those specific skills that are required.
  - Decide who will be on the crew.

  In many companies, the project manager or job superintendent determines the size and makeup of the crew. Then, supervision expects the crew leader to accomplish the goals and objectives with the crew provided. Even though the crew leader may not have any involvement in staffing the crew, the crew leader is responsible for training the crew members to ensure that they have the skills needed to do the job. In addition, the crew leader is responsible for keeping the crew adequately staffed at all times to avoid job delays. This involves dealing with absenteeism and turnover, two common problems discussed in earlier sections.

4.2.4 Scheduling

Planning and scheduling are closely related and are both very important to a successful job. Planning identifies the required activities and how and in what order to complete them. Scheduling involves establishing start and finish times/dates for each activity.

A schedule for a project typically shows the following:

- Operations listed in sequential order
- Units of construction
- Duration of activities
- Estimated date to start and complete each activity
- Quantity of materials to be installed

There are different types of schedules used today. They include the bar chart, the network schedule, also called the critical path method (CPM) or precedence diagram, and the short-term, or look-ahead schedule.

The following is a summary of the steps a crew leader must complete to develop a schedule.

Step 1 Make a list of all the activities that the plan requires to complete the job, including individual work activities and special tasks, such as inspections or the delivery of materials. At this point, the crew leader should just be concerned with generating a list, not with determining how to accomplish the activities, who will perform them, how long they will take, or the necessary sequence to complete them.

Step 2 Use the list of activities created in Step 1 to reorganize the work activities into a logical sequence. When doing this, keep in mind that certain steps can't happen before the completion of others. For example, footing excavation must occur before concrete emplacement.

Step 3 Assign a duration or length of time that it will take to complete each activity and determine the start time for each. Then place each activity into a schedule format. This step is important because it helps the crew leader compare the task time estimates to the scheduled completion date or time.

The crew leader must be able to read and interpret the job schedule. On some jobs, the form provides the beginning and expected end date for each activity, along with the expected crew or worker's production rate. The crew leader can use this information to plan work more effectively, set realistic goals, and compare the starts and completions of tasks to those on the schedule.

Before starting a job, the crew leader must do the following:

- Determine the materials, tools, equipment, and labor needed to complete the job.
- Determine when the various resources are needed.
- Follow up to ensure that the resources are available on the job site when needed.

The crew leader should verify the availability of needed resources three to four working days before the start of the job. This should occur even earlier for larger jobs. Advance preparation will help avoid situations that could potentially delay starting the job or cause it to fall behind schedule.

Supervisors can use bar chart schedules, also known as *Gantt charts*, for both short-term and long-term jobs. However, they are especially helpful for jobs of short duration.

Bar charts provide management with the following:

- A visual presentation of the overall time required to complete the job using a logical method rather than a calculated guess
- A means to review the start and duration of each part of the job
- Timely coordination requirements between crafts
- Alternative sequences of performing the work

A bar chart works as a control device to see whether the job is on schedule. If the job isn't on schedule, supervision can take immediate action in the office and the field to correct the problem and increases the likelihood of completing the activity on schedule. *Figure 27* illustrates a bar or Gantt chart.

Another type of schedule is the network schedule, which shows dependent (critical path) activities and other activities completed in parallel with but not part of the critical path. In *Figure 28*, for example, reinforcing steel can't be set until the concrete forms have been built and placed. Other activities are happening in parallel, but the forms are in the critical path.

When building a house, builders can't install and finish drywall until wiring, plumbing, and HVAC ductwork have been roughed-in. Because other activities, such as painting and trim work, depend on drywall completion, the drywall work is a critical-path operation. In other words, until it is complete, workers can't start the other tasks, and the project itself will likely experience delay by the amount of delay starting any dependent

Did You Know?

Recycling Materials

Habitat for Humanity, a charitable organization that builds homes for disadvantaged families, accepts surplus building materials for use in their projects. In some cities, they have stores called ReStores, which serve as retail outlets for such materials. They may also accept materials salvaged during demolition of a structure. Companies can obtain LEED credits through practices such as salvaging building materials, segregating scrap materials for recycling, and taking steps to minimize waste.

Figure Credit: Sushil Shenoy/Virginia Tech

Did You Know?

Scheduling Apps

Many companies use project scheduling software. Among the top construction project management applications at the time of publication are WorkflowMax, Procore, BuildTools, Co-Construct, and BuilderTREND. Such programs are spreadsheets on steroids, which can respond to schedule changes by altering downstream events and producing a revised schedule. Most of the newest programs are cloud-based, and users can access and update them remotely using mobile, tablet, and computer devices.

activity. Likewise, workers can't even start the drywall installation until the rough-ins are complete. Therefore, the project superintendent is likely to focus on those activities when evaluating schedule performance.

The advantage of a network schedule is that it allows project leaders to see how a schedule change with one activity is likely to affect other activities and the project in general. Planners lay out a network schedule on a timeline and usually show the estimated duration for each activity. Planners use network schedules for complex jobs that take a long time to complete. The PERT (program evaluation and review technique) schedule is a form of network schedule.

Since the crew needs to hold to the job schedule, the crew leader needs to be able to plan daily production. As discussed earlier, short-term scheduling is a method used to do this. *Figure 29* displays an example of a short-term, look-ahead schedule.

The information to support short-term scheduling comes from the estimate or cost breakdown. The schedule helps to translate estimate data and the various job plans into a day-to-day schedule of events. The short-term schedule provides the crew leader with visibility over the immediate future of the project. If actual production begins to slip behind estimated production, the schedule will warn the crew leader that a problem lies ahead and that a schedule slippage is developing.

Crew leaders should use short-term scheduling to set production goals. Generally, workers can improve production when they:

- Know the amount of work to be accomplished
- Know the time available to complete the work
- Can provide input when setting goals

Example:

A carpentry crew on a retaining wall project is about to form and pour catch basins and put up wall forms. The crew has put in several catch basins, so the crew leader is sure that they can perform the work within the estimate. However, the crew leader is concerned about their production of the wall forms. The crew will work on both the basins and the wall forms at the same time. The scheduling process in this scenario could be handled as follows:

1. The crew leader notices the following in the estimate or cost breakdown:
   a. Production factor for wall forms = 16 worker-hours (w-h) per 100 ft^2
   b. Work to be done by measurement = 800 ft^2
   c. Total time: (800 ft^2 × 16 w-h/100 ft^2) = 128 w-h
2. The carpenter crew consists of the following:
   a. One carpenter crew leader
   b. Four carpenters
   c. One laborer
3. The crew leader determines the goal for the job should be set at 128 w-h (from the cost breakdown).
4. If the crew remains the same (six workers), the work should be completed in about 21 crew-hours (128 w-h ÷ 6 workers/crew = 21.3 crew-hours).
5. The crew leader then discusses the production goal (completing 800 ft^2 in 21 crew-hours) with the crew and encourages them to work together to meet the goal of getting the forms erected within the estimated time.

In this example, the crew leader used the short-term schedule to translate production into work-hours or crew-hours and to schedule work so that the crew can accomplish it within the estimate. In addition, setting production targets provides the motivation to produce more than the estimate requires.

No matter what type of schedule is used, supervision must keep it up to date to be useful to the crew leader. Inaccurate schedules are of no value. The person responsible for scheduling in the office handles the updates. This person uses information gathered from job field reports to do the updates.

The crew leader is usually not directly involved in updating schedules. However, completing field or progress reports used by the company may be a daily responsibility to keep the schedule up to date. It's critical that the crew leader fill out any required forms or reports completely and accurately.

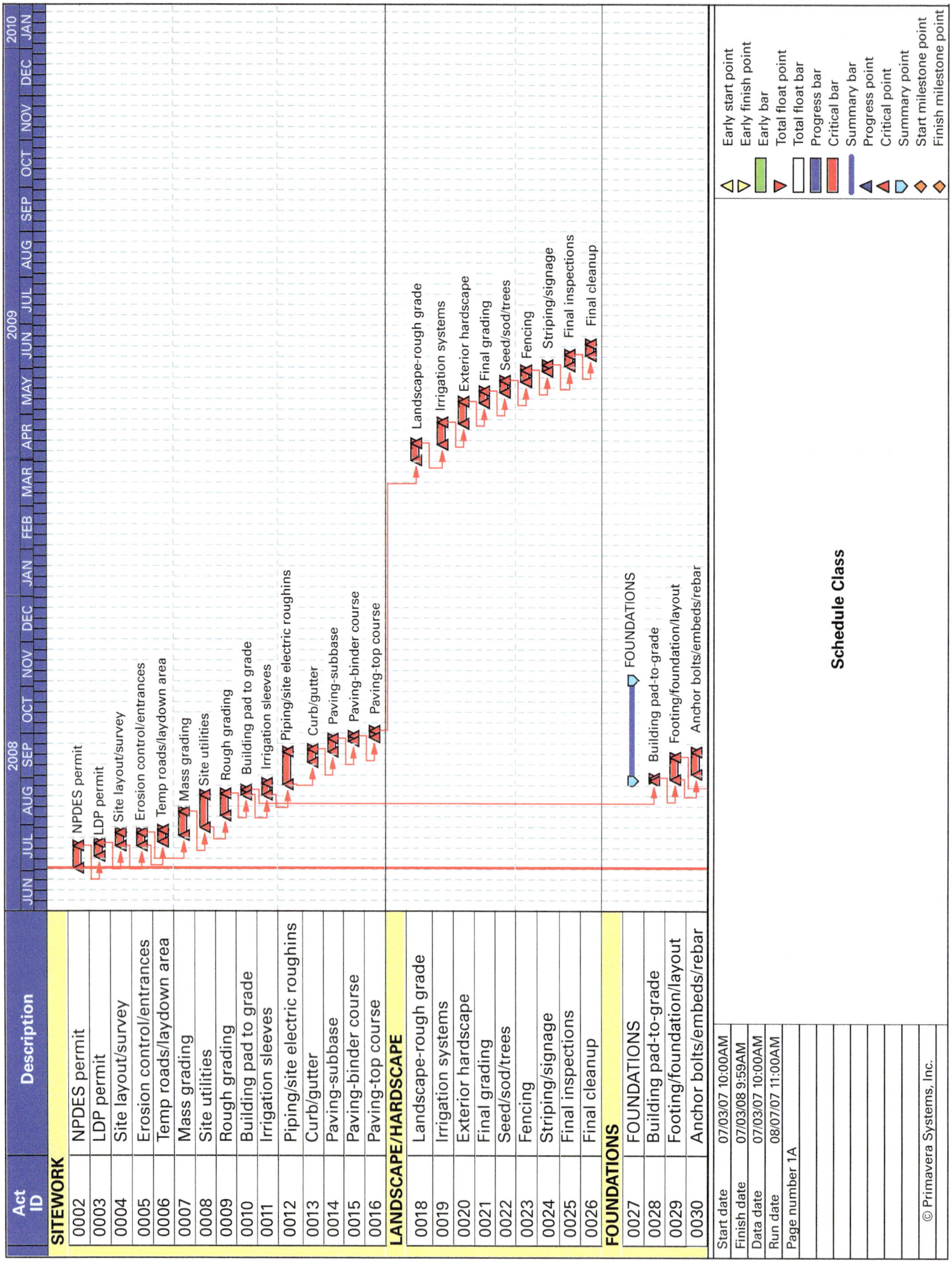

Figure 27 Example of a bar-chart schedule.

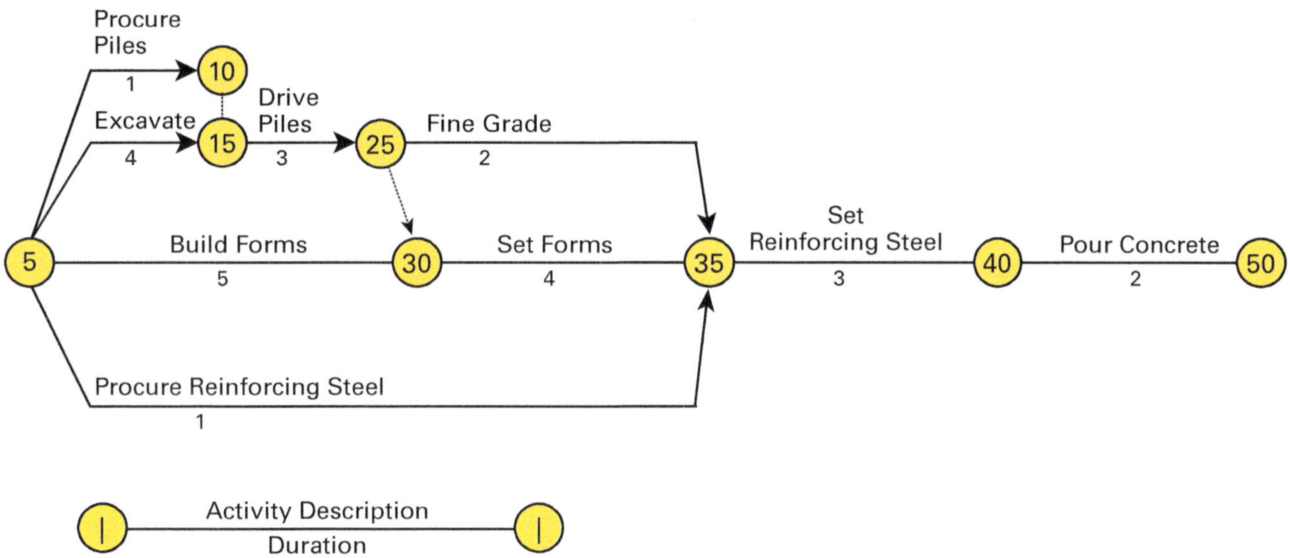

Figure 28 Example of a network schedule.

4.3.0 Cost Control

Being aware of costs and controlling them is the responsibility of every employee on the job. It's the crew leader's job to ensure that employees uphold this responsibility. Control refers to the comparison of estimated performance against actual performance and following up with any needed corrective action. Crew leaders who use cost-control practices are more valuable to the company than those who do not.

On a typical job, many activities are going on at the same time, even within a given crew. This can make it difficult to control the activities involved. The crew leader must be constantly aware of the costs of a project and effectively control the various resources used on the job.

When resources aren't controlled, the cost of the job increases. For example, a plumbing crew of four people is installing soil pipe and runs out of fittings. Three crew members wait (*Figure 30*) while one crew member goes to the plumbing-supply dealer for a part that costs only a few dollars. It takes the crew member an hour to get the part, so four worker-hours of production have been lost. In addition, the total cost of the delay must include the travel costs for retrieving the supplies.

4.3.1 Assessing Cost Performance

Cost performance on a project is determined by comparing actual costs to estimated costs. Regardless of whether the job is a contract-bid project or an in-house project, the company must first establish a budget. In the case of a contract bid, the budget is generally the cost estimate used to bid the job. For an in-house job, participants will submit labor and material forecasts, and someone in authority will authorize a project budget.

It is common to estimate cost by either breaking the job into funded tasks or by forecasting labor and materials expenditures on a timeline. Many companies create a **work breakdown structure (WBS)** for each project. Within the WBS, planners assign each major task a discrete charge number. Anyone working on that task charges that number on their time sheet, so that project managers can readily track cost performance. However, knowing how much money the company is spending doesn't necessarily determine cost performance.

Although financial reports can show that actual expenses are tracking with forecast expenses, they don't show if the work itself is occurring at the required rate. Thus, it is possible to have spent half the budget, but have less than half of the work compete. When the project is broken down into funded tasks related to schedule activities and events, there is far greater control over cost performance.

4.3.2 Field Reporting System

The total estimated cost comes from the job estimates, but managers obtain the actual cost of doing the work from an effective field-reporting system. A field-reporting system consists of a series of forms, which are completed by the crew leader and others. Each company has its own forms and methods for obtaining information. The following paragraphs describe the general information and the process of how they are used. First, records must document the number of hours each person worked on each task. This

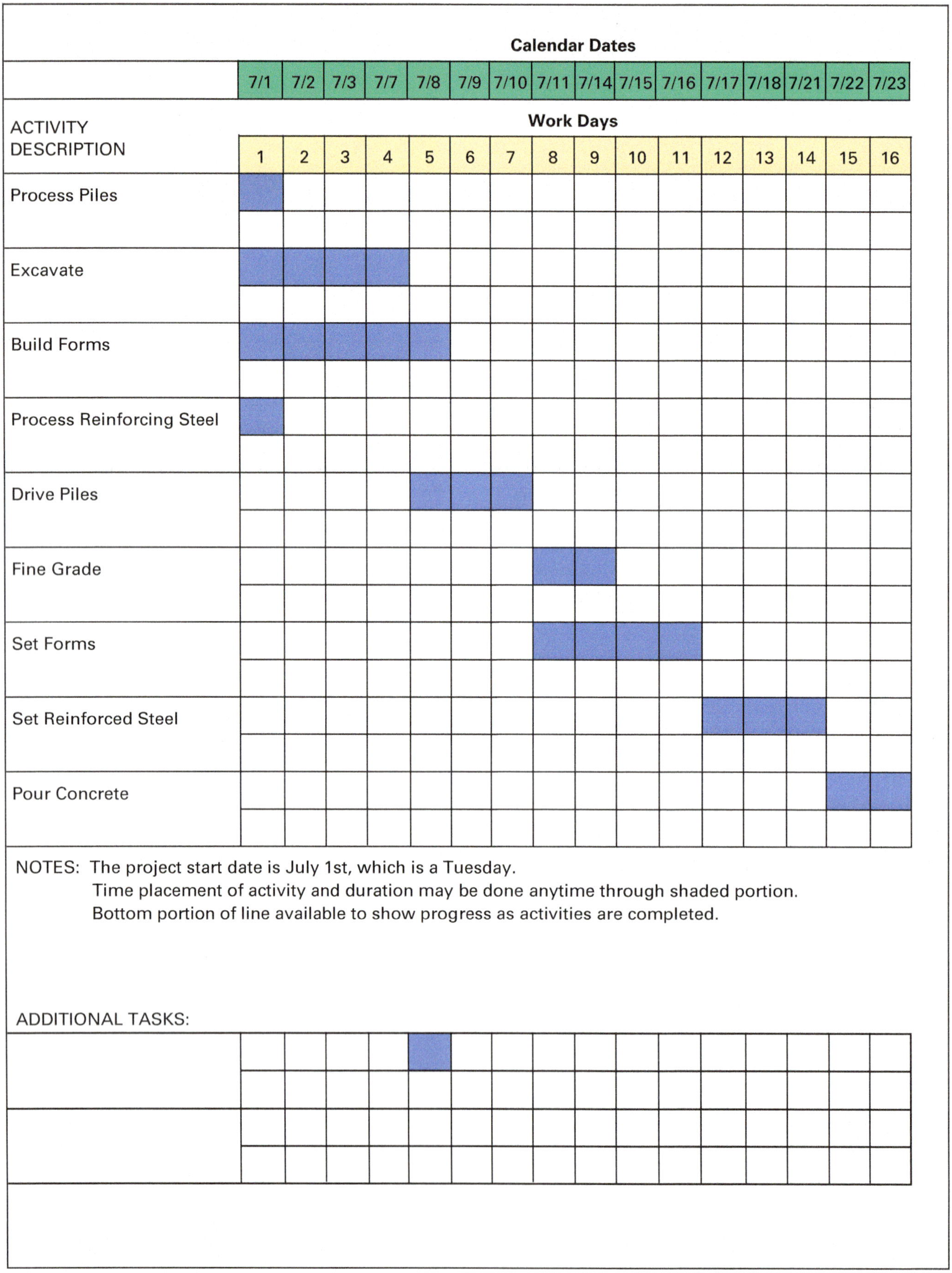

Figure 29 Short-term schedule.

Figure 30 Idle workers cost money.

information comes from daily time cards. Once the accounting department knows how many hours each employee worked on an activity, it can calculate the total cost of the labor by multiplying the number of hours worked by the wage rate for each worker. Managers can then calculate the cost for the labor to do each task as the job progresses. They compare this cost and the estimated costs during the project and at its completion. Managers use a similar process to determine if the costs to operate equipment are comparable to the estimated equipment operating costs.

As the crew places materials, a designated person will measure the quantities from time to time, and send this information to the company office and, possibly, the crew leader. This information, along with the actual cost of the material and the number of hours it took the workers to install it, is compared to the estimated cost. If the cost is greater than the estimate, management and the crew leader must take action to reduce the cost.

For this comparison process to be of use, the information obtained from field personnel must be correct. It is important that the crew leader be accurate in reporting. This is another reason to maintain a daily job diary as discussed earlier. In the event of a legal/contractual conflict with the client, courts can use such diaries as evidence in legal proceedings, and they can be helpful in reaching a settlement.

Example:

You are running a crew of five concrete finishers for a subcontractor. When you and your crew show up to finish a slab, the general contractor (GC) informs you, "We're a day behind on setting the forms, so I need you and your crew to stand down until tomorrow." What do you do?

Solution:

You should first call your office to let them know about the delay. Then, immediately record it in your job diary. A six-man crew for one day represents 48 worker-hours. If your company charges $30 an hour, that's a potential loss of $1,440, which the company would want to recover from the GC. If there is a dispute, your entry in the job diary could result in a favorable decision for your employer.

4.3.3 Crew Leader's Role in Cost Control

The crew leader is often the company's representative in the field, where the work takes place. Therefore, the crew leader contributes a great deal to determining job costs. When supervision assigns work to a crew, the crew leader should receive a budget and schedule for completing the job. It is then up to the crew leader to make sure the crew finishes the job on time and stays within budget. The crew leader achieves this by actively managing the use of labor, materials, tools, and equipment.

"Beware of little expenses;
a small leak will sink a great ship."
—Benjamin Franklin

If the actual costs are at or below the estimated costs, the job is progressing as planned and scheduled, then the company will realize the expected profit. However, if the actual costs exceed the estimated costs, one or more problems may result in the company losing its expected profit, and maybe more. No company can remain in business if it continually loses money.

One of the factors that can increase cost is client-related changes (often called scope creep). The crew leaders must be able to assess the potential impact of such changes and, if necessary, confer with the employer to determine the course of action. If contractor-related losses are occurring, the crew leader and superintendent will need to work together to get the costs back in line.

"There is no such thing as scope creep, only scope gallop."
—Cornelius Fitchner

The following are some causes that can make actual costs exceed estimated costs, along with some potential actions the crew leader can take to bring the costs under control:

> **NOTE**
>
> Before starting any action, the crew leader should check with company supervision to see that the action proposed is acceptable and within the company's policies and procedures.
>
> - *Late delivery of materials, tools, and/or equipment* – Plan ahead to ensure that job resources will be available when needed
> - *Inclement weather* – Work with the superintendent and have alternate plans ready
> - *Unmotivated workers* – Counsel the workers
> - *Accidents* – Enforce the existing safety program (or evaluate the program for adequacy)

There are many other methods to get the job done on time if it gets off schedule. Examples include working overtime, increasing the size of the crew, prefabricating assemblies, or working staggered shifts. However, these examples may increase the cost of the job, so the crew leader should not do them without the approval of the project manager.

4.4.0 Resource Control

The crew leader's job is to complete assigned tasks safely according to the plans and specifications, on schedule, and within the scope of the estimate. To accomplish this, the crew leader must closely control how resources of materials, equipment, tools, and labor are used. The crew must minimize waste whenever possible.

Control involves measuring performance and correcting departures from plans and specifications to accomplish objectives. Control anticipates predictable deviations from plans and specifications based on experience and takes measures to prevent them from occurring.

An effective control process can be broken down into the following steps:

Step 1 Establish standards and divide them into measurable units. For example, a baseline can be created using experience gained on a typical job, where 2,000 linear feet (LF) of 1¼" copper water tube was installed in five days. Thus, dividing the total LF installed by the number of installation days gives the average installation rate. In this case, for 1¼" copper water tube, the standard rate was 400 LF/day.

Step 2 Measure performance against a standard. On another job, a crew placed 300 LF of the same tube during a given day. Thus, this actual production of 300 LF/day did not meet the standard rate of 400 LF/day.

Step 3 Adjust operations to meet the standard. In Step 2, if the plan scheduled the job for five days, the crew leader would have to take action to meet this goal. If 300 LF/day is the actual average daily production rate for that job, the crew must increase their placement rate by 100 LF/day to meet the standard. This increase may come by making the workers place piping faster, adding workers to the crew, or receiving authorization for overtime.

The crew leader's responsibility in materials control depends on the policies and procedures of the company. In general, the crew leader is responsible for ensuring on-time delivery, preventing waste, controlling delivery and storage, and preventing theft of materials.

4.4.1 Ensuring On-Time Delivery

It's essential that the materials required for each day's work be on the job site when needed. The crew leader should confirm in advance the placement of orders for all materials and that they will arrive on schedule. A week or so before the delivery date, follow up to make sure there will be no delayed deliveries or items on backorder.

If other people are responsible for providing the materials for a job, the crew leader must follow up to make sure that the materials are available when needed. Otherwise, delays occur as crew members stand around waiting for the delivery of materials.

4.4.2 Preventing Waste

Waste in construction can add up to loss of critical and costly materials and may result in job delays. The crew leader needs to ensure that every crew member knows how to use the materials efficiently. Crew leaders should monitor their crews to make certain that no materials are wasted.

An example of waste is a carpenter who saws off a piece of lumber from a full-sized piece, when workers could have found the length needed in the lumber scrap pile. Another example of waste

Did You Know?

Just in Time (JIT)

Just-in-time (JIT) delivery is a strategy in which materials arrive at the job site when needed. This means that a crew may install the materials right off the truck. This method reduces the need for on-site storage and staging areas. It also reduces the risk of loss or damage while moving products about the site. Other modern material management methods include the use of radiofrequency identification (RFID) tags that make it easy to inventory and locate material in crowded staging areas. Using an RFID inventory control system requires special scanners.

involves rework due to bad technique or a conflict in craft scheduling requiring removal and reinstallation of finished work. The waste in time spent and replacement material costs occur when redoing the task a second time.

Under LEED, waste control is very important. Companies receive LEED credits for finding ways to reduce waste and for recycling construction waste. Workers should segregate waste materials by type for recycling, if feasible (*Figure 31*).

4.4.3 Verifying Material Delivery

A crew leader may be responsible for the receipt of materials delivered to the work site. When this happens, the crew leader should require a copy of the shipping invoice or similar document and check each item on the invoice against the actual materials to verify delivery of the correct amounts.

The crew leader should also check the condition of the materials to verify that nothing is defective before signing the invoice. This can be difficult and time consuming because it requires the crew to open cartons and examine their contents. However, this step cannot be overlooked, because a signed invoice indicates that the recipient accepted all listed materials and in an undamaged state. If the crew leader signs for the materials without checking them, and then finds damage, no one will be able to prove that the materials came to the site in that condition.

After checking and signing the shipping invoice, the crew leader should give the original or a copy to the superintendent or project manager. The company then files the invoice for future reference because it serves as the only record the company has to check bills received from the supplier company.

4.4.4 Controlling Delivery and Storage

Another essential element of materials control is the selection of where the materials will be stored on the job site. There are two factors in determining the appropriate storage location. The first is

Figure 31 Waste materials separated for recycling.

convenience. If possible, the materials should be stored near where the crew will use them. The time and effort saved by not having to carry the materials long distances will greatly reduce the installation costs.

Next, the materials must be stored in a secure area to avoid damage. It is important that the storage area suit the materials being stored. For instance, materials that are sensitive to temperature, such as chemicals or paints, should be stored in climate-controlled areas to prevent waste.

4.4.5 Preventing Theft and Vandalism

Theft and vandalism of construction materials increase costs. The direct and indirect costs of lost materials include obtaining replacement materials, production time lost while the needed materials are missing, and the costs of additional security precautions can add significantly to the overall cost to the company. In addition, the insurance that the contractor purchases will increase in cost as the theft and vandalism rate grows.

The best way to avoid theft and vandalism is a secure job site. At the end of each work day, store unused materials and tools in a secure location, such as a locked construction trailer. If the job site is fenced or the building has lockable accesses, the materials can be stored within. Many sites have security cameras and/or intrusion alarms to help minimize theft and vandalism.

4.4.6 Equipment Control

The crew leader may not be responsible for long-term equipment control. However, the equipment required for a specific job is often the crew leader's responsibility. The first step is to schedule when a crew member must transport the required equipment from the shop or rental yard. The crew leader is responsible for informing the shop of the location of its use and seeing a worker returns it to the shop when finished with it.

It is common for equipment to lay idle at a job site because of a lack of proper planning for the job and the equipment arrived early. For example, if wire-pulling equipment arrives at a job site before the conduit is in place, this equipment will be out of service while awaiting the conduit installation. In addition to the wasted rental cost, damage, loss, or theft can also occur with idle equipment while awaiting use.

The crew leader needs to control equipment use, ensure that the crew operates the equipment according to its instructions, and within time and cost guidelines. The crew leader must also assign or coordinate responsibility for maintaining and repairing equipment as indicated by the applicable preventive maintenance schedule. Delaying maintenance and repairs can lead to costly equipment failures. The crew leader must also ensure that the equipment operators have the necessary credentials to operate the equipment, including applicable licenses.

The crew leader is responsible for the proper operation of all other equipment resources, including cars and trucks. Reckless or unsafe operation of vehicles will likely result in damaged equipment, leading to disciplinary action in the case of the workers, fines and repair costs levied by the vehicle vendors, and a loss of confidence in the crew leader's ability to govern crew-members' actions.

The crew leader should also arrange to secure all equipment at the close of each day's work to prevent theft. If continued use of the equipment is necessary for the job, the crew leader should make sure to lock it in a safe place; otherwise, return it to the shop.

4.4.7 Tool Control

Among companies, various policies govern who provides hand and power tools to employees. Some companies provide all the tools, while others furnish only the larger power tools. The crew leader should be familiar with and enforce any company policies related to tools.

Tool control has two aspects. First, the crew leader must control the issue, use, and maintenance of all tools provided by the company. Second, the crew leader must control how crew members use the tools to do the job. This applies to tools supplied by the company as well as tools that belong to the workers.

Using the proper tools correctly saves time and energy. In addition, proper tool use reduces the chance of damage to the tool during use, as well as injury to the user and to nearby workers. This is especially true of edge tools and those with any form of fixed or rotating blade. A dull tool is far more dangerous than a sharp one.

Tools must be adequately maintained and properly stored. Making sure that tools are cleaned, dried, and lubricated prevents rust and ensures that the tools are in the proper working order. If a tool is damaged, it is essential to repair or replace it promptly. Otherwise, an accident or injury could result.

> **NOTE**
> Regardless of whether a worker or the company owns a tool, OSHA holds companies responsible for the consequences of using a tool on a job site. The company is accountable if an employee receives an injury from a defective tool. Therefore, the crew leader needs to be aware of any defects in the tools the crew members are using.

Users should take care of company-issued tools as if they were their own. Workers should not abuse tools simply because they don't their property.

One of the major sources of low productivity on a job is the time spent searching for a tool. To prevent this from occurring, supervision should establish a storage location for company-issued tools and equipment. The crew leader should make sure that crew members return all company-issued tools and equipment to this designated location after use. Similarly, workers should organize their personal toolboxes so that they can readily find the appropriate tools and return their tools to their toolboxes when they are finished using them. This is a matter of professionalism that crew leaders are in an excellent position to develop.

Studies have shown that some keys to an effective tool control system are:

- Limiting the number of people allowed access to stored tools
- Limiting the number of people held responsible for tools and holding them accountable
- Controlling the ways in which a tool can be returned to storage
- Making sure tools are available when needed

4.4.8 Labor Control

Labor typically represents more than half the cost of a project, and therefore has an enormous impact on profitability. For that reason, it's essential to manage a crew and their work environment in a way that maximizes their productivity. One of the ways to do that is to minimize the time spent on unproductive activities such as the following:

- Engaging in bull sessions
- Correcting drawing errors
- Retrieving tools, equipment, and materials
- Waiting for other workers to finish

If crew members are habitually goofing off, it is up to the crew leader to counsel those workers. The crew leader's daily diary should document the counseling. The crew leader will need to refer repeated violations to the attention of higher management as guided by company policy.

All sorts of errors will occur during a project that need correction. Some errors, such as mistakes on drawings, may be outside of the crew leader's control. However, before a drawing error results in performing an action that will need to be corrected later, a careful examination of the drawings by the crew leader may discover the error before work begins, saving time and materials. If crew members are making mistakes due to inexperience, the crew leader can help avoid these errors by providing on-the-spot training and by checking on inexperienced workers more often.

The availability and location of tools, equipment, and materials to a crew can have a profound effect on their productivity. As discussed in the previous section, the key to minimizing such problems is proactive management of these resources.

Delays caused by crews or contractors can be minimized or avoided by carefully tracking the project schedule. In doing so, crew leaders can anticipate delays that will affect the work and either take action to prevent the delay or redirect the crew to another task.

4.4.9 Production and Productivity

Production is the amount of construction materials put in place. It is the quantity of materials installed on a job, such as 1,000 linear feet of waste pipe installed for a task. On the other hand, productivity is a rate, and depends on the level of efficiency of the work. It is the amount of work done per hour or day by one worker or a crew.

Production levels are set during the estimating stage. The estimator determines the total amount of materials placed from the plans and specifications. After the job is complete, supervision can assess the actual amount of materials installed, and can compare the actual production to the estimated production.

Productivity relates to the amount of materials put in place by the crew over a certain period of time. The estimator uses company records during the estimating stage to determine how much time and labor it will take to place a certain quantity of materials. From this information, the estimator calculates the productivity necessary to complete the job on time.

Participant Exercise G

1. List the methods your company uses to minimize waste.

____

____

____

____

____

2. List the methods your company uses to control small tools on the job.

____

____

____

____

____

3. List five ways that you feel your company could control labor to maximize productivity.

____

____

____

____

____

For example, it might take a crew of two people ten days to paint 5,000 square feet. To calculate the required average productivity, divide 5,000 ft^2 by 10 days. The result is 500 ft^2/day. The crew leader can compare the daily production of any crew of two painters doing similar work with this average, as discussed previously.

Planning is essential to productivity. The crew must be available to perform the work and have all the required materials, tools, and equipment in place when the job begins.

The time on the job should be for business, not for taking care of personal problems. Anything not work-related should be handled after hours, away from the job site, if possible. Even the planning of after-work activities, arranging social functions, or running personal errands should occur after work or during breaks. Only very limited and necessary exceptions to these rules should be permitted (e.g., making or meeting medical appointments), and these should be spelled out in company or crew policies.

Organizing field work can save time. The key to effectively using time is to work smarter, not necessarily harder. Working smart can save your crew from doing unnecessary work. For example, most construction projects require that the contractor submit a set of as-built plans at the completion of the work. These plans describe the actual structure and installation of materials. The best way to prepare these plans is to mark up a set of working plans as the work is in progress as described earlier. That way, workers and supervision won't forget pertinent details and waste time trying to reconstruct how the company did the work.

"Knowledge is power is time is money."
—Robert Thier, Storm and Silence

The amount of material used should not exceed the estimated amount. If it does, either the estimator has made a mistake, undocumented changes have occurred, or rework has caused the need for additional materials. Whatever the case, the crew leader should use effective control techniques to ensure the efficient use of materials.

When bidding a job, most companies calculate the cost per worker-hour. For example, a ten-day job might convert to 160 w-h (two painters for ten days at eight hours per day). If the company charges a labor rate of $30/hour, the labor cost would be $4,800. The estimator then adds the cost of materials, equipment, and tools, along with overhead costs and a profit factor, to determine the price of the job.

After completion of a job, information gathered through field reporting allows the company office to calculate actual productivity and compare it to the estimated figures. This helps to identify productivity issues and improves the accuracy of future estimates.

The following labor-related practices can help to ensure productivity:

- Ensure that all workers have the required resources when needed.
- Ensure that all personnel know where to go and what to do after each task is completed.
- Make reassignments as needed.
- Ensure that all workers have completed their work properly.

Additional Resources

Construction Workforce Development Professional, NCCER. 2016. New York, NY: Pearson Education, Inc.

Mentoring for Craft Professionals, NCCER. 2016. New York, NY: Pearson Education, Inc.

Blueprint Reading for Construction, James A. S. Fatzinger. 2003. New York, NY: Pearson Education, Inc.

Construction Leadership from A to Z: 26 Words to Lead By, Wally Adamchik. 2011. Live Oak Book Company.

The following websites offer resources for products and training:

Architecture, Engineering, and Construction Industry (AEC), **www.aecinfo.com**

US Green Building Council (USGBC), **www.usgbc.org/leed**

4.0.0 Section Review

1. Which of these activities occurs during the development phase of a project?
   a. Architect/engineer sketches are prepared and a preliminary budget is developed.
   b. Government agencies give a final inspection of the design, check for adherence to codes, and inspect materials used.
   c. Detailed project drawings and specifications are prepared.
   d. Contracts for the project are awarded.

2. A job diary should typically record any of the following *except* _____.
   a. items such as job interruptions and visits
   b. the exact times of scheduled lunch breaks
   c. changes needed to project drawings
   d. the actual time each job task related to a particular project took to complete

3. Which of the following is a correct statement regarding project cost?
   a. Cost is handled by the accounting department and isn't a concern of the crew leader.
   b. The difference between the estimated cost and the actual cost affects a company's profit.
   c. Wasted material is factored into the estimate and is never a concern.
   d. The contractor's overhead costs aren't included in the cost estimate.

4. To prevent job delays due to late delivery of materials, the crew leader should _____.
   a. demand a discount from the supplier to compensate for the delay
   b. tell a crew member to go look for the delivery truck
   c. refuse the late delivery and re-order the materials from another supplier
   d. check with the supplier in advance of the scheduled delivery

SUMMARY

In construction, the crew leader is the company's supervisor on the ground at the scene where the actual work takes place. The modern crew leader must be prepared to deal with a host of challenges ranging from differences in age, culture, race, gender, language, and educational backgrounds to attitudes regarding authority and personal integrity. Rather than being overwhelmed, the effective and successful crew leader will use these challenges to broaden his or her understanding of people and to grow in character.

Crew leaders must develop an understanding of the company, its policies, and how to become an effective and valuable member of the supervisory team. They must also be able to effectively communicate up and down the chain, to motivate the crew's workers, and develop a sense of ownership within the crew so that they have the confidence and will to do the jobs.

Safety is an overriding concern for the company, and crew leaders need to make safe work practices and conditions the number-one priority. Once a worker has lost an eye, a limb, or life itself, it's too late to reconsider how the task might have been done safer. Workers are a company's most valuable asset and the crew leader is in the position of being the best example of safety to the crew.

While crew leaders view their role as mainly people and work managers, these tasks gain a greater sense of importance when crew leaders understand their responsibilities to the company for whom they work. Effectively controlling worker productivity, resources, schedules, and other aspects of the project directly contribute to cost control, which reflects in greater profits for the company. Companies recognize effective crew leaders and they can expect long and successful careers as a result.

Review Questions

1. Younger-generation workers may have a better perception of success if _____.
   a. crew leaders assign their projects in smaller, well-defined units
   b. they receive closer supervision
   c. crew leaders leave them alone
   d. they are given the perception that they are in charge

2. Companies can help prevent sexual harassment in the workplace by _____.
   a. requiring employee training that avoids the potentially offensive subject of stereotypes
   b. developing a consistent policy with appropriate consequences for engaging in sexual harassment
   c. communicating to workers that, legally, the victim of sexual harassment is only the one being directly harassed
   d. establishing an effective complaint or grievance process for victims of harassment

3. Which of the following statements about a formal organization is *true*?
   a. It is a group of members created mainly to share professional information.
   b. It is a group established for developing industry standards.
   c. It is a group of individuals led by someone all working toward the same goal.
   d. It is a group that is identified by its overall function rather than by key positions.

4. Which of these departments in a large company would likely arrange equipment rentals?
   a. Accounting
   b. Estimating
   c. Human Resources
   d. Purchasing

5. When a person is promoted to crew leader, the amount of time they spend on craft work _____.
   a. increases
   b. decreases
   c. stays the same
   d. doubles

6. Which of the following is a *correct* statement regarding annual crew evaluations?
   a. Their results should never come as a surprise to the crew member.
   b. Formal evaluations should be conducted every three months.
   c. Comments on performance should be withheld until the formal evaluation.
   d. Advise the crew members to improve their performance just before the formal evaluation is scheduled.

7. As a crew leader, what should you avoid when giving recognition to crew members?
   a. Looking for good work and knowing what good work is
   b. Being available at the job site to witness good work
   c. Giving praise for poor work to make the worker feel better
   d. Encouraging work improvement through your confidence in a worker

8. Within the construction industry, who normally assigns a crew to the crew leader?
   a. The US Department of Labor
   b. The company
   c. The crew leader
   d. The crew members

9. Some decisions are difficult to make because you _____.
   a. have no experience dealing with the circumstances requiring a decision
   b. don't want to make a decision
   c. prefer to think it's someone else's responsibility to make a decision
   d. don't recognize the need for a decision

10. The first step in problem solving should be _____.
    a. choosing a solution to the problem
    b. evaluating the solution to the problem
    c. defining the problem
    d. testing the solution to the problem

11. The leading cause of fatalities in the construction industry is _____.
    a. electrocution
    b. asphyxiation
    c. struck by something
    d. falls

12. OSHA inspection of a business or job site _____.
    a. can be done only by invitation
    b. can be conducted at random
    c. is done only after an accident
    d. is conducted only if a safety violation occurs

13. To proactively identify and assess hazards, it is essential that employers _____.
    a. ensure that crew leaders are present and available on the job site
    b. conduct safety inspections of contractor offices
    c. design safety logs to document safety violations
    d. conduct accident investigations

14. Employers who are subject to OSHA recordkeeping requirements must maintain a log of recordable occupational injuries and illnesses for _____.
    a. three years
    b. five years
    c. seven years
    d. as long as the company is in business

15. Which of the following statements regarding safety training sessions is *correct*?
    a. The project manager usually holds them.
    b. They are held only for new employees.
    c. They should be conducted frequently by the crew leader.
    d. They are required only after an accident has occurred.

16. The type of contract in which the client pays the contractor for their actual labor and material expenses they incur is known as a _____.
    a. firm fixed-price contract
    b. time-spent contract
    c. performance-based contract
    d. cost-reimbursable contract

17. On a design-build project, _____.
    a. the same contractor is responsible for both design and construction
    b. the owner is responsible for providing the design
    c. the architect does the design and the general contractor builds the project
    d. a construction manager is hired to oversee the project

18. When planning labor for onsite work, the crew leader must _____.
    a. factor in the absenteeism and turnover rates
    b. factor in time for training classes
    c. select the material to be used by each crew member
    d. identify the skills needed to perform the work

19. Which of the following is *not* a source of data in the field-reporting system?
    a. The cost-estimate for the project
    b. Crew leader's daily diary
    c. Onsite inventory of placed materials
    d. Daily time cards

20. The following duties are important in work site equipment control *except* _____.
    a. scheduling the pickup and return of a piece of equipment
    b. ensuring the equipment is operated and maintained correctly
    c. monitoring safe and responsible operation of vehicles
    d. selection and purchase of equipment to be stocked at the work site equipment room

Trade Terms Introduced in This Module

Autonomy: The condition of having complete control over one's actions, and being free from the control of another. To be independent.

Bias: A preconceived inclination against or in favor of something.

Cloud-based applications: Mobile and desktop digital programs that can connect to files and data located in distributed storage locations on the Internet ("the Cloud"). Such applications make it possible for authorized users to create, edit, and distribute content from any location where Internet access is possible.

Craft professionals: Workers who are properly trained and work in a particular construction trade or craft.

Crew leader: The immediate supervisor of a crew or team of craft professionals and other assigned persons.

Critical path: In manufacturing, construction, and other types of creative processes, the required sequence of tasks that directly controls the ultimate completion date of the project.

Demographics: Social characteristics and other factors, such as language, economics, education, culture, and age, that define a statistical group of individuals. An individual can be a member of more than one demographic.

Ethics: The moral principles that guides an individual's or organization's actions when dealing with others. Also refers to the study of moral principles.

Infer: To reach a conclusion using a method of reasoning that starts with an assumption and considers a set of logically-related events, conditions, or statements.

Intangible: Not touchable, material, or measureable; lacking a physical presence.

Job description: A description of the scope and responsibilities of a worker's job so that the individual and others understand what the job entails.

Job diary: A written record that a supervisor maintains periodically (usually daily) of the events, communications, observations, and decisions made during the course of a project.

Legend: In maps, plans, and diagrams, an explanatory table defining all symbolic information contained in the document.

Lethargy: Sluggishness, slow motion, lack of activity, or a lack of enthusiasm.

Letter of instruction (LOI): A written communication from a supervisor to a subordinate that informs the latter of some inadequacy in the individual's performance, and provides a list of actions that the individual must satisfactorily complete to remediate the problem. Usually a key step in a series of disciplinary actions.

Local area networks (LAN): Communication networks that link computers, printers, and servers within a small, defined location, such as a building or office, via hard-wired or wireless connections.

Lockout/tagout (LOTO): A system of safety procedures for securing electrical and mechanical equipment during repairs or construction, consisting of warning tags and physical locking or restraining devices applied to controls to prevent the accidental or purposeful operation of the equipment, endangering workers, equipment, or facilities.

Look-ahead schedule: A manual- or software-scheduling tool that looks several weeks into the future at the planned project events; used for anticipating material, labor, tool, and other resource requirements, as well as identifying potential schedule conflicts or other problems.

Negligence: Lack of appropriate care when doing something, or the failure to do something, usually resulting in injury to an individual or damage to equipment.

Organizational chart: A diagram that shows how the various management and operational responsibilities relate to each other within an organization. Named positions appear in ranked levels, top to bottom, from those functional units with the most and broadest authority to those with the least, with lines connecting positions indicating chains of authority and other relationships.

Paraphrase: Rewording a written or verbal statement in one's own words in such a way that the intent of the original statement is retained.

Pragmatic: Sensible, practical, and realistic.

Proactive: To anticipate and take action in the present to deal with potential future events or outcomes based on what one knows about current events or conditions.

Project manager: In construction, the individual who has overall responsibility for one or more construction projects; also called the general superintendent.

Return on investment (ROI): A measure of the gain or loss of money resulting from an investment; normally measured as a percentage of the original investment.

Safety data sheets (SDS): Documents listing information about a material or substance that includes common and proper names, chemical composition, physical forms and properties, hazards, flammability, handling, and emergency response in accordance with national and international hazard communication standards. Also called material safety data sheets (MSDS).

Sexual harassment: Any unwelcome verbal or nonverbal form of communication or action construed by an individual to be of a sexual or gender-related nature.

Smartphone: The most common type of cellular telephone that combines many other digital functions within a single mobile device. Most important of these, besides the wireless telephone, are high-resolution still and video cameras, text and email messaging, and GPS-enabled features.

Superintendent: In construction, the individual who is the on-site supervisor in charge of a given construction project.

Synergy: Any type of cooperation between organizations, individuals, or other entities where the combined effect is greater than the sum of the individual efforts.

Textspeak: A form of written language characteristic of text messaging on mobile devices and text-based social media, which usually consists of acronyms, abbreviations, and minimal punctuation.

Wide area networks (WAN): Dedicated communication networks of computers and related hardware that serve a given geographic area, such as a work site, campus, city, or a larger but distinct area. Connectivity is by wired and wireless means, and may use the Internet as well.

Wi-Fi: The technology allowing communications via radio signals over a LAN or WAN equipped with a wireless access point or the Internet. (Wi-Fi stands for wireless fidelity.) Many types of mobile, portable, and desktop devices can communicate via Wi-Fi connections.

Work breakdown structure (WBS): A diagrammatic and conceptual method for subdividing a complex project, concept, or other thing into its various functional and organizational parts so that planners can analyze and plan for each part in detail.

OSHA Forms for Safety Records

Employers who are subject to the recordkeeping requirements of the Occupational Safety and Health Act of 1970 must maintain records of all recordable occupational injuries and illnesses. The following are three important OSHA forms used for this recordkeeping (these forms can be accessed at **www.osha.gov**):

- OSHA Form 300, *Log of Work-Related Injuries and Illnesses*
- OSHA Form 300A, *Summary of Work-Related Injuries and Illnesses*
- OSHA Form 301, *Injury and Illness Incident Report*

OSHA's Form 300 (Rev. 01/2004)

Log of Work-Related Injuries and Illnesses

Attention: This form contains information relating to employee health and must be used in a manner that protects the confidentiality of employees to the extent possible while the information is being used for occupational safety and health purposes.

Year ________

U.S. Department of Labor
Occupational Safety and Health Administration

You must record information about every work-related injury or illness that involves loss of consciousness, restricted work activity or job transfer, days away from work, or medical treatment beyond first aid. You must also record significant work-related injuries and illnesses that are diagnosed by a physician or licensed health care professional. You must also record work-related injuries and illnesses that meet any of the specific recording criteria listed in 29 CFR 1904.8 through 1904.12. Feel free to use two lines for a single case if you need to. You must complete an injury and illness incident report (OSHA Form 301) or equivalent form for each injury or illness recorded on this form. If you're not sure whether a case is recordable, call your local OSHA office for help.

Form approved OMB no. 1218-0176

Establishment name ____________________

City ____________ State ____________

Identify the person			Describe the case			Classify the case										
						CHECK ONLY ONE box for each case based on the most serious outcome for that case:			Enter the number of days the injured or ill worker was:		Check the "injury" column or choose one type of illness: (M)					
						Days away from work	Remained at work		Away From Work (days)	On job transfer or restriction (days)						
(A) Case No.	(B) Employee's Name	(C) Job Title (e.g., Welder)	(D) Date of injury or onset of illness (mo./day)	(E) Where the event occurred (e.g. Loading dock north end)	(F) Describe injury or illness, parts of body affected, and object/substance that directly injured or made person ill (e.g. Second degree burns on right forearm from acetylene torch)	(H)	Job transfer or restriction (I)	Other record-able cases (J)	(K)	(L)	Injury (1)	Skin Disorder (2)	Respiratory Condition (3)	Poisoning (4)	Hearing Loss (5)	All other illnesses (6)
					Page totals	0	0	0	0	0	0	0	0	0	0	0
											Injury (1)	Skin Disorder (2)	Respiratory Condition (3)	Poisoning (4)	Hearing Loss (5)	All other illnesses (6)

Be sure to transfer these totals to the Summary page (Form 300A) before you post it.

Public reporting burden for this collection of information is estimated to average 14 minutes per response, including time to review the instruction, search and gather the data needed, and complete and review the collection of information. Persons are not required to respond to the collection of information unless it displays a currently valid OMB control number. If you have any comments about these estimates or any aspects of this data collection, contact: US Department of Labor, OSHA Office of Statistics, Room N-3644, 200 Constitution Ave, NW, Washington, DC 20210. Do not send the completed forms to this office.

Page 1 of 1

Figure A01 OSHA Form 300

Figure A02 OSHA Form 300A

Year ______

U.S. Department of Labor
Occupational Safety and Health Administration

Form approved OMB no. 1218-0176

OSHA's Form 300A (Rev. 01/2004)

Summary of Work-Related Injuries and Illnesses

All establishments covered by Part 1904 must complete this Summary page, even if no injuries or illnesses occurred during the year. Remember to review the Log to verify that the entries are complete and accurate before

Using the Log, count the individual entries you made for each category. Then write the totals below, making sure you've added the entries from every page of the log. If you had no cases write "0."

Employees former employees, and their representatives have the right to review the OSHA Form 300 in its entirety. They also have limited access to the OSHA Form 301 or its equivalent. See 29 CFR 1904.35, in OSHA's Recordkeeping rule, for further details on the access provisions for these forms.

Number of Cases

Total number of deaths	Total number of cases with days away from work	Total number of cases with job transfer or restriction	Total number of other recordable cases
0	0	0	0
(G)	(H)	(I)	(J)

Number of Days

Total number of days away from work	Total number of days of job transfer or restriction
0	0
(K)	(L)

Injury and Illness Types

Total number of...
(M)

(1) Injury	0	(4) Poisoning	0
(2) Skin Disorder	0	(5) Hearing Loss	0
(3) Respiratory Condition	0	(6) All Other Illnesses	0

Post this Summary page from February 1 to April 30 of the year following the year covered by the form

Public reporting burden for this collection of information is estimated to average 58 minutes per response, including time to review the instruction, search and gather the data needed, and complete and review the collection of information. Persons are not required to respond to the collection of information unless it displays a currently valid OMB control number. If you have any comments about these estimates or any aspects of this data collection, contact: US Department of Labor, OSHA Office of Statistics, Room N-3644, 200 Constitution Ave, NW, Washington, DC 20210. Do not send the completed forms to this office.

Establishment information

Your establishment name ______________________

Street ______________________

City ____________ State ____________ Zip ______

Industry description (e.g., Manufacture of motor truck trailers)

____

Standard Industrial Classification (SIC), if known (e.g., SIC 3715)

__ __ __ __

OR North American Industrial Classification (NAICS), if known (e.g., 336212)

__ __ __ __ __ __

Employment information

Annual average number of employees ______

Total hours worked by all employees last year ______

Sign here

Knowingly falsifying this document may result in a fine.

I certify that I have examined this document and that to the best of my knowledge the entries are true, accurate, and complete.

______________ Company executive ______ Title

______________ Phone ______ Date

OSHA's Form 301
Injuries and Illnesses Incident Report

Attention: This form contains information relating to employee health and must be used in a manner that protects the confidentiality of employees to the extent possible while the information is being used for occupational safety and health purposes.

U.S. Department of Labor
Occupational Safety and Health Administration

Form approved OMB no. 1218-0176

This *Injury and Illness Incident Report* is one of the first forms you must fill out when a recordable work-related injury or illness has occurred. Together with the *Log of Work-Related injuries and Illnesses* and the accompanying *Summary*, these forms help the employer and OSHA develop a picture of the extent and severity of work-related incidents.

Within 7 calendar days after you receive information that a recordable work-related injury or illness has occurred, you must fill out this form or an equivalent. Some state workers' compensation, insurance, or other reports may be acceptable substitutes. To be considered an equivalent form, any substitute must contain all the information asked for on this form.

According to Public Law 91-596 and 29 CFR 1904, OSHA's recordkeeping rule, you must keep this form on file for 5 years following the year to which it pertains.

If you need additional copies of this form, you may photocopy and use as many as you need.

Completed by ______

Title ______

Phone ______ Date ______

Information about the employee

1) Full Name ______

2) Street ______

City ______ State ______ Zip ______

3) Date of birth ______

4) Date hired ______

5) ☐ Male
☐ Female

Information about the physician or other health care professional

6) Name of physician or other health care professional ______

7) If treatment was given away from the worksite, where was it given?

Facility ______

Street ______

City ______ State ______ Zip ______

8) Was employee treated in an emergency room?
☐ Yes
☐ No

9) Was employee hospitalized overnight as an in-patient?
☐ Yes
☐ No

Information about the case

10) Case number from the Log ______ *(Transfer the case number from the Log after you record the case.)*

11) Date of injury or illness ______

12) Time employee began work ______ AM/PM

13) Time of event ______ AM/PM ☐ Check if time cannot be determined

14) **What was the employee doing just before the incident occurred?** Describe the activity, as well as the tools, equipment or material the employee was using. Be specific. Examples: "climbing a ladder while carrying roofing materials"; "spraying chlorine from hand sprayer"; "daily computer key-entry."

15) **What happened?** Tell us how the injury occurred. Examples: "When ladder slipped on wet floor, worker fell 20 feet"; "Worker was sprayed with chlorine when gasket broke during replacement"; "Worker developed soreness in wrist over time."

16) **What was the injury or illness?** Tell us the part of the body that was affected and how it was affected; be more specific than "hurt", "pain", or "sore." Examples: "strained back"; "chemical burn, hand"; "carpal tunnel syndrome."

17) **What object or substance directly harmed the employee?** Examples: "concrete floor"; "chlorine"; "radial arm saw." If this question does not apply to the incident, leave it blank.

18) **If the employee died, when did death occur?** Date of death

Public reporting burden for this collection of information is estimated to average 22 minutes per response, including time for reviewing instructions, searching existing data sources, gathering and maintaining the data needed, and completing and reviewing the collection of information. Persons are not required to respond to the collection of information unless it displays a current valid OMB control number. If you have any comments about this estimate or any other aspects of this data collection, including suggestions for reducing this burden, contact: US Department of Labor, OSHA Office of Statistics, Room N-3644, 200 Constitution Ave, NW, Washington, DC 20210. Do not send the completed forms to this office.

Figure A03 OSHA Form 301

Additional Resources

This module presents thorough resources for task training. The following reference material is recommended for further study.

Construction Workforce Development Professional, NCCER. 2016. New York, NY: Pearson Education, Inc.

Mentoring for Craft Professionals, NCCER. 2016. New York, NY: Pearson Education, Inc.

Blueprint Reading for Construction, James A. S. Fatzinger. 2003. New York, NY: Pearson Education, Inc.

Construction Leadership from A to Z: 26 Words to Lead By, Wally Adamchik. 2011. Live Oak Book Company.

Generational Cohorts and their Attitudes Toward Work Related Issues in Central Kentucky, Frank Fletcher, et al. 2009. Midway College, Midway, KY. **www.kentucky.com.**

It's Your Ship: Management Techniques from the Best Damn Ship in the Navy, Captain D. Michael Abrashoff, USN. 2012. New York City, NY: Grand Central Publishing.

Survival of the Fittest, Mark Breslin. 2005. McNally International Press.

The Definitive Book of Body Language: The Hidden Meaning Behind People's Gestures and Expressions, Barbara Pease and Allan Pease. 2006. New York City, NY: Random House / Bantam Books.

The Young Person's Guide to Wisdom, Power, and Life Success: Making Smart Choices, Brian Gahran, PhD. 2014. San Diego, CA: Young Persons Press. **www.WPGBlog.com**.

The following websites offer resources for products and training:

Aging Workforce News, **www.agingworkforcenews.com**.

American Society for Training and Development (ASTD), **www.astd.org**.

Architecture, Engineering, and Construction Industry (AEC), **www.aecinfo.com**.

Equal Employment Opportunity Commission (EEOC), **www.eeoc.gov**.

National Association of Women in Construction (NAWIC), **www.nawic.org**.

National Census of Fatal Occupational Injuries (NCFOI), **www.bls.gov**.

National Institute of Occupational Safety and Health (NIOSH), **www.cdc.gov/niosh**.

National Safety Council, **www.nsc.org**.

Occupational Safety and Health Administration (OSHA), **www.osha.gov**.

Society for Human Resources Management (SHRM), **www.shrm.org**.

United States Census Bureau, **www.census.gov**.

United States Department of Labor, **www.dol.gov**.

US Green Building Council (USGBC), **www.usgbc.org/leed**.

Wi-Fi® is a registered trademark of the Wi-Fi Alliance, **www.wi-fi.org**

Figure Credits

© Photographerlondon/Dreamstime.com, Module Opener
© iStockphoto.com/shironosov, Figure 1
© iStockphoto.com/kali9, Figures 2, 10
© iStockphoto.com/nanmulti, Figure 3
© iStockphoto.com/Halfpoint, Figure 13
© iStock.com/vejaa, Figure 16
DIYawards, Figure 17
© iStockphoto.com/Tashi-Delek, Figure 18
© iStockphoto.com/kemaltaner, Figure 23
Sushil Shenoy/Virginia Tech, Figure 31
John Ambrosia, Figure 27
© iStockphoto.com/izustun, Figure 30

Section Review Answer Key

Answer	Section Reference	Objective
Section One		
1. a	1.1.0	1a
2. b	1.2.2	1b
3. b	1.3.0	1c
Section Two		
1. b	2.0.0	2a
2. c	2.2.1	2b
3. a	2.3.1	2c
4. c	2.4.1	2d
5. a	2.5.1	2e
Section Three		
1. a	3.0.0	3a
2. d	3.2.0; Table 1	3b
3. a	3.3.0	3c
4. b	3.4.0	3d
Section Four		
1. a	4.1.1	4a
2. b	4.2.2	4b
3. b	4.3.3	4c
4. d	4.4.1	4d

NCCER CURRICULA — USER UPDATE

NCCER makes every effort to keep its textbooks up-to-date and free of technical errors. We appreciate your help in this process. If you find an error, a typographical mistake, or an inaccuracy in NCCER's curricula, please fill out this form (or a photocopy), or complete the online form at **www.nccer.org/olf**. Be sure to include the exact module ID number, page number, a detailed description, and your recommended correction. Your input will be brought to the attention of the Authoring Team. Thank you for your assistance.

Instructors – If you have an idea for improving this textbook, or have found that additional materials were necessary to teach this module effectively, please let us know so that we may present your suggestions to the Authoring Team.

NCCER Product Development and Revision
13614 Progress Blvd., Alachua, FL 32615
Email: curriculum@nccer.org
Online: www.nccer.org/olf

❑ Trainee Guide ❑ Lesson Plans ❑ Exam ❑ PowerPoints Other ______________

Craft / Level: Copyright Date:

Module ID Number / Title:

Section Number(s):

Description:

Recommended Correction:

Your Name:

Address:

Email: Phone:

This page is intentionally left blank.

Index

A

ABC. *See* Associated Builders and Contractors (ABC)
Absenteeism, (46101):28–29
Accidents
  cell phones and, (46101):4
  costs of
    direct and indirect, (46101):33
    insured and uninsured, (46101):33–34
  investigations, (46101):37–38, 41
  OSHA notification requirements, (46101):39
  preventing through inspection, (46101):39
  root causes of, (46101):38
  statistics, (46101):32
Accomplishment, motivation and, (46101):2, 20, 22
Accountability, (46101):9
Accounting department, (46101):8
Active listening, (46101):18–19
ADA. *See* Americans with Disabilities Act (ADA)
Advancement opportunities, motivating with, (46101):22–23
Advisor style of leadership, (46101):15, 16
African-American Vernacular English, (46101):6
AGC. *See* Associated General Contractors (AGC)
Alcohol abuse, (46101):40–41
American Society for Training and Development (ASTD), (46101):3
Americans with Disabilities Act (ADA), (46101):11
Apprenticeship programs, (46101):2
As-built drawings, (46101):48
Associated Builders and Contractors (ABC), (46101):8
Associated General Contractors (AGC), (46101):8
ASTD. *See* American Society for Training and Development (ASTD)
Attitude, worker, (46101):16
Authority
  defined, (46101):9
  delegating, (46101):9
  leadership and, (46101):15–16
  respect and, (46101):15–16
Autonomy, (46101):13, 15, 78
Awards, safety-related, (46101):42–43

B

Baby Boomers, (46101):2, 5
Bar chart schedule, (46101):62, 64, 66
Bias, (46101):13, 14, 78
Bid price, (46101):51
Budgeting in construction planning, (46101):48, 51–54
Business ethics, (46101):16
Business organization
  authority and accountability in, (46101):9
  division of responsibility in, (46101):8–9
  formal vs. informal, (46101):8
  job descriptions, (46101):9–10, 11
  organizational charts, (46101):8, 9
  policies and procedures, (46101):10
  typical departments, (46101):8

C

Cellular telephones, (46101):3–4
Change, motivating with, (46101):22
Cloud-based applications, (46101):1, 4, 78
Communication
  barriers, overcoming, (46101):18, 20
  in diverse workforces, (46101):5–6
  effective, (46101):20
  electronic, (46101):4
  feedback in, (46101):3, 18–19
  language barriers, (46101):5–6
  leadership and, (46101):14
  nonverbal, (46101):19–20
  orders, (46101):26
  organizational structure for, (46101):8
  in project planning, (46101):59
  promoting safety through, (46101):41
  tailoring, (46101):20
  verbal, (46101):17–19
  visual, (46101):19
  written, (46101):19
Communication process
  the message in the, (46101):18
  the receiver in the, (46101):18
  the sender in the, (46101):17–18
  steps in the, (46101):17
Communication styles, gender differences in, (46101):5
Community-corporate partnerships, (46101):2
Conflict
  cultural, resolving, (46101):6
  employee inability to work with others, (46101):28
Construction costs
  controlling
    cost performance assessment and, (46101):65
    crew leader's role, (46101):67–68
    field reporting system for, (46101):65, 67
  direct and indirect, (46101):48, 51
  estimating and budgeting, (46101):48, 51–54
  resource controls
    equipment, (46101):70
    labor, (46101):71
    materials, (46101):69–70
    production and productivity, (46101):71–73
    steps in, (46101):68
    tools, (46101):70–71
    waste prevention, (46101):68–69
Construction industry, present-day
  diversity initiatives, (46101):5
  gender and cultural issues, (46101):5–8
  technology's impact, (46101):3–4
  training's role in, (46101):2
  workforce shortages, (46101):2
Construction management
  communication in, (46101):59
  delivery systems
    construction management, (46101):48, 50
    design-build, (46101):48, 50
    general contracting, (46101):48, 50
  planning
    elements in, (46101):54–55
    goals, establishing, (46101):58
    process steps, (46101):58–60
    stages of, (46101):55–56

Construction management
planning (*continued*)
tasks, identifying, (46101):58–59
project flow
communicating responsibilities, (46101):58
construction phase, (46101):48
development phase, (46101):46–47
follow-up activities, (46101):59–60
overview, (46101):49
planning phase, (46101):47–48
project-delivery systems, (46101):48
required work, identifying the, (46101):57–58
resource planning
equipment, (46101):61
labor, (46101):61
materials, (46101):60
safety, (46101):60
tools, (46101):61
scheduling
bar chart schedule, (46101):62, 64, 66
network schedule, (46101):62–63, 65
process, (46101):61–62
short-term, (46101):62–63
updating, (46101):63
site planning, key elements, (46101):61
Construction management delivery systems, (46101):48, 50
Contests, safety-related, (46101):41
Contracts
administration of, (46101):8
cost reimbursable, (46101):47–48
firm fixed price, (46101):47
guaranteed maximum price (GMP), (46101):48
Controller style of leadership, (46101):15
Corporate-community partnerships, (46101):2
Cost estimating, (46101):48, 51–54
Cost performance assessment, (46101):65
Cost reimbursable contracts, (46101):47–48
Craft professionals, (46101):1, 2, 78
Craft training, (46101):3
Craftworkers vs. crew leaders, (46101):13
Crew leaders
authority and accountability in, (46101):9, 15–16
craftworkers vs., (46101):13
defined, (46101):1, 78
leadership transition, (46101):13–14
managerial vs. non-managerial time, (46101):14
present day challenges, (46101):1
requirements of, (46101):13–14
responsibilities of
controlling construction costs, (46101):67–68
delegating, (46101):25–26
orders, issuing, (46101):26
policies and procedures, implementing and enforcing, (46101):25–26, 29
Critical path, (46101):46, 61, 78
Cultural issues in the workplace
conflicts, resolving, (46101):6
language barriers, (46101):5–6

D

Deaths, construction industry, (46101):33
Decision making
leadership and, (46101):15
problem solving vs., (46101):27
Decisions, types of, (46101):27
Delegating, (46101):9, 25–26
Delivery systems
design-build, (46101):48, 50
general contracting, (46101):47, 50
Demographics, (46101):1, 2, 78
Department of Labor (DOL), (46101):2
Design-build delivery systems, (46101):48, 50
Directive style of leadership, (46101):15–16
Disability discrimination, (46101):11
Disciplinary action, (46101):29
Discrimination in the workplace, (46101):7–8, 11
Diversity, workplace, (46101):5–8
DOL. *See* Department of Labor (DOL)
Drawings
as-built drawings, (46101):48
construction phase, (46101):48
redline, (46101):48
Drug abuse, (46101):40–41

E

Ebonics, (46101):6
EEOC. *See* Equal Employment Opportunity Commission (EEOC)
Electronic communication, (46101):4
English as a second language training, (46101):6
Equal Employment Opportunity Commission (EEOC), (46101):7
Equipment, controlling as resource, (46101):70
Estimating in construction planning, (46101):48, 51–54
Estimating office, (46101):8
Ethics, (46101):13, 16–17, 26, 78
Executives, (46101):8

F

Failure, leadership and, (46101):14
Fall fatalities, (46101):32
Feedback
in communication, (46101):3, 18–19
for motivation, (46101):3
training and, (46101):3
Field reporting system, (46101):65, 67
Fire protection and prevention, (46101):40
Firm fixed price contracts, (46101):47
First aid, (46101):39–40

G

Gantt charts, (46101):62
Gender issues in the workplace
communication, (46101):5
discrimination based on gender, (46101):7
sexual harassment, (46101):6–7
General contracting delivery systems, (46101):48, 50
Generation X, (46101):5
Generation Y. *See* Millenials
Generation Z, (46101):5
Goal setting
motivation and, (46101):2, 3
in planning projects, (46101):58
successful teams and, (46101):24
Guaranteed maximum price contracts, (46101):48

H

Habitat for Humanity, (46101):62
Hazard identification and assessment, (46101):36
Human Resources Department, (46101):8, 40

I
Incentives, safety-related, (46101):42–43
Infer, (46101):13, 20, 78
Injury statistics, work-related, (46101):32
Inspections, construction phase, (46101):48
Intangible, (46101):32, 33, 78

J
Job descriptions, (46101):1, 9–10, 11, 78
Job diary, (46101):46, 78
Job importance, motivation and, (46101):22
Job satisfaction, (46101):20
Just-in-time delivery, (46101):69

L
Labor control, (46101):71
Labor planning, (46101):61
Labor shortages, (46101):2
LAN. *See* Local area networks (LAN)
Language barriers, (46101):5–6
Leaders
  characteristics of effective, (46101):14–15
  defined, (46101):14
  ethical conduct in, (46101):16–17
  expectations of, (46101):15
  functions of, (46101):15
Leadership in Energy and Environmental Design (LEED), (46101):50
Leadership styles
  advisor, (46101):15, 16
  controller, (46101):15
  directive, (46101):15–16
Learning styles, (46101):2, 3, 4
LEED. *See* Leadership in Energy and Environmental Design (LEED)
Legal ethics, (46101):16
Legend, (46101):13, 19, 69, 78
Lethargy, (46101):32, 40, 78
Letter of instruction (LOI), (46101):1, 7, 78
Listening, (46101):18–19
Local area networks (LAN), (46101):1, 4, 78
Lockout/tagout (LOTO), (46101):13, 29, 78
LOI. *See* Letter of instruction (LOI)
Look-ahead schedule, (46101):46, 58, 78
LOTO. *See* Lockout/tagout (LOTO)

M
Materials control
  delivery verification, (46101):69
  just-in-time delivery, (46101):69
  storage, (46101):69–71
  theft and vandalism prevention, (46101):70
  on-time delivery, (46101):68
  waste prevention, (46101):68
Materials planning, (46101):60
Message element in communication, (46101):18
Millennials, (46101):5
Minority discrimination, (46101):7–8
Morale, (46101):20
Motivation, employee
  inferring, (46101):20
  leadership and, (46101):14
  safety-related, (46101):41
  tips for crew leaders, (46101):24
Motivators
  accomplishment, (46101):20, 22
  advancement opportunity, (46101):22–23
  change, (46101):22
  feedback, (46101):3
  financial, (46101):2
  goal setting, (46101):2, 3
  incentives and awards, (46101):42–43
  job importance, (46101):22
  personal growth, (46101):23
  recognition and praise, (46101):23
  rewards, (46101):23–24
  sense of achievement, (46101):2

N
National Association of Women in Construction (NAWIC), (46101):8
National Institute of Occupational Safety and Health (NIOSH), (46101):33
NAWIC. *See* National Association of Women in Construction (NAWIC)
Negligence, (46101):32, 33, 78
Network schedule, (46101):62–63, 65
NIOSH. *See* National Institute of Occupational Safety and Health (NIOSH)
Nonverbal communication, (46101):19–20

O
Occupational Safety and Health Act of 1970, (46101):34
Occupational Safety and Health Administration (OSHA)
  accident notification requirements, (46101):39
  fire safety standards, (46101):40
  first aid training requirements, (46101):39–40
  Form 300, *Log of Work-Related Injuries and Illnesses*, (46101):81
  Form 300A, *Summary of Work-Related Injuries and Illnesses*, (46101):82
  Form 301, *Injury and Illness Incident Report*, (46101):41, 83
  function of, (46101):34–35
  penalties for violations, (46101):35
  regulation enforcement, (46101):35
  safety policies and procedures regulations, (46101):35–36
  on-site tools regulations, (46101):71
  state run programs, (46101):35
  workplace inspections, (46101):35
Operations office, (46101):8
Orders, issuing, (46101):26
Organizational charts, (46101):1, 8, 9, 78
Organizational skills, (46101):15
OSHA. *See* Occupational Safety and Health Administration (OSHA)

P
Paraphrase, (46101):1, 6, 18–19, 78
Performance evaluations, (46101):19
Persistence, leadership and, (46101):14
Personal growth, motivation and, (46101):23
Policies and procedures
  defined, (46101):10
  enforcing, (46101):29
  failure to comply with, (46101):29
  function, (46101):10
  implementing, (46101):25–26
  safety and health programs, (46101):35–36
Pragmatic, (46101):5, 78
Praise, motivating with, (46101):23
Proactive, (46101):13, 29, 79
Problem solving
  decision making vs., (46101):27
  process, (46101):27–28
Productivity, (46101):71–73

Professional ethics, (46101):16
Project manager, (46101):13, 79
Publicity, safety-related, (46101):41
Purchasing department, (46101):8

Q

Quantity takeoff, (46101):51, 52

R

Receiver of communication, (46101):18
Recognition, motivating with, (46101):23
Record systems, safety, (46101):36–37
Redline drawings, (46101):48
Resource controls
  equipment, (46101):70
  labor, (46101):71
  materials, (46101):69–70
  production and productivity, (46101):71–73
  steps in, (46101):68
  tools, (46101):70–71
  waste prevention, (46101):68–69
Resource planning
  equipment, (46101):61
  labor, (46101):61
  materials, (46101):60
  safety, (46101):60
  site, (46101):61
  tools, (46101):61
Respect, (46101):16
Respectful workplace training, (46101):7
Responsibility
  defined, (46101):9
  ownership of, (46101):9
ReStores, (46101):62
Return on investment (ROI), (46101):46, 47, 79
Rewards, motivating with, (46101):23–24
ROI. *See* Return on investment (ROI)
Rural White Southern English, (46101):6

S

Safety
  costs of
    accidents, (46101):33–34
    for regulatory non-compliance, (46101):35
  crew leader involvement
    accident investigation, (46101):41
    fire protection and prevention, (46101):40
    first aid, (46101):39–40
    inspections, (46101):39
    promotion, (46101):41
    substance abuse, (46101):40–41
    training sessions, (46101):38–39
  employer responsibilities
    accident investigation, (46101):37–38
    hazard identification and assessment, (46101):36
    information and training, (46101):36
    policies and procedures, (46101):35–36
    record systems, (46101):36–37
    safety programs, (46101):35
  promoting
    communication in, (46101):41
    contests, (46101):42
    incentives and awards, (46101):42–43
    publicity for, (46101):42–43
Safety data sheets (SDS), (46101):32, 36, 79
Safety hazards
  the fatal four, (46101):32
  identification and assessment, (46101):36
Safety planning, (46101):60
Safety policies and procedures
  for accident investigation, (46101):37–38
  employer responsibilities, (46101):35–36
  hazard identification and assessment, (46101):36
  information and training on, (46101):36
  record systems requirements of, (46101):36–37
Scheduling, project
  bar chart schedule, (46101):62, 64, 66
  network schedule, (46101):62–63, 65
  process, (46101):61–62
  short-term, (46101):62–63
  updating, (46101):63
SDS. *See* Safety data sheets (SDS)
Sender communicator, (46101):17–18
Sensitivity training, (46101):7
Sexual harassment, (46101):1, 5, 6–7, 79
Short-term scheduling, (46101):62–63
SHRM. *See* Society of Human Resources Management (SHRM)
Site planning, key elements, (46101):61
Situational ethics, (46101):16
Smartphone, (46101):1, 3, 79
Society of Human Resources Management (SHRM), (46101):3
Substance abuse, (46101):40–41
Superintendent, (46101):13, 79
Synergy, (46101):1, 6, 79

T

Teams
  building successful, (46101):24–25
  delegating within, (46101):25–26
Technology, impact on industry, (46101):3–4
Textspeak, (46101):13, 19, 79
Theft, materials, (46101):70
Tool control, (46101):70–71
Tool planning, (46101):61
Training
  communication and, (46101):3
  craft training, (46101):3
  English as a second language, (46101):6
  feedback in, (46101):3
  in-house classes, (46101):3
  impact of technology on, (46101):3–4
  lack of, reasons for, (46101):3
  learning styles and, (46101):2, 3, 4
  need for, (46101):2
  respectful, (46101):7
  respectful workplace classes, (46101):7
  safety-related, (46101):35, 36, 38–39
  SHRM tips on, (46101):3
  supervisory, (46101):3
  tips for supervisors in charge of, (46101):3
Turnover, (46101):29

V

Vandalism, preventing, (46101):70
Visual communication, (46101):19

W

WAN. *See* Wide area network (WAN)
Waste prevention, (46101):68–69
WBS. *See* Work breakdown structure (WBS)
Wide area network (WAN), (46101):1, 4, 79
Wi-Fi, (46101):1, 4, 79

Women in industry
communication styles, (46101):5
sexual harassment of, (46101):5, 6–7
Work breakdown structure (WBS), (46101):46, 75, 79
Workers
common problems
absenteeism, (46101):28–29
inability to work with others, (46101):28
policies and procedures, failure to comply, (46101):29
substance abuse, (46101):40–41
turnover, (46101):29
disciplining, (46101):29
motivators
accomplishment, (46101):20, 22
advancement opportunity, (46101):22–23
change, (46101):22
feedback, (46101):3
financial, (46101):2
goal setting, (46101):2, 3
job importance, (46101):22
personal growth, (46101):23
recognition and praise, (46101):23
rewards, (46101):23–24
sense of achievement, (46101):2
Workers compensation, (46101):33
Workforce
diversity, (46101):5
generational differences, (46101):2, 4–5
shortages predicted, (46101):2
Written communication, (46101):19

This page is intentionally left blank.

This page is intentionally left blank.

This page is intentionally left blank.